SPITBALLS FROM THE BACK ROW

To CATHERINE -

ESSAYS ON THE MODERN AMERICAN EDUCATION SYSTEM

A new friend in writing - and in the spirit ☺

Hope you enjoy -

Camyn Campbell

AUG 2020

SPITBALLS FROM THE BACK ROW

ESSAYS ON THE MODERN AMERICAN EDUCATION SYSTEM

LARRY CAMPBELL

an imprint of
OGHMA CREATIVE MEDIA

Meath Press
An imprint of Oghma Creative Media, Inc.
2401 Beth Lane, Bentonville, Arkansas 72712

Names: Cambell, Larry, author.
Title: Spitballs from the Back Row/Larry Campbell
Description: First Edition. | Bentonville: Meath, 2018.
Identifiers: ISBN: 978-1-63373-481-4 (hardcover) |
ISBN: 978-1-63373-468-5 (trade paperback) | ISBN: 978-1-63373-482-1 (eBook)
Subjects: BISAC: EDUCATION / Philosophy, Theory & Social Aspects |
EDU008000 EDUCATION / Decision-Making & Problem Solving |
MAT030000 MATHEMATICS / Study & Teaching

Meath Press trade paperback edition October, 2018

Cover Design by Casey W. Cowan
Interior Design by Erin Ladd
Editing by Michael L. Frizell, Jeremy Menefee, Cyndy Prasse Miller, and Amy Cowan

DEDICATION

This book is dedicated to **Pat**, my wife of *almost* a half-century. Whatever level of quality these former columns have reached, that level would not have been nearly as high without her.

In my very early days of column writing, I used to share my columns with Pat before turning them in. I suspect I was secretly hoping for what I thought was the obligatory wifely praise and confirmation. Instead, she gave me something much better: her honest opinion! As a writer (and a life partner) this used to mildly irritate me. Then I began to listen to her comments.

I quickly discovered that her comments (which was, often to say, her questions) were insightful—not necessarily from the viewpoint of a "editor," but from the more important viewpoint of a "reader." It became crystal clear to me that if *she*—knowing me as well as anyone—wasn't sure what I was saying, then I needed to re-write it.

It turns out that there are two things that most help a babbling-professor-trying-to-be-a-columnist. One is a 600-word limit and the other is a spouse who is willing to express her honest opinion. Thank heavens I had both.

ACKNOWLEDGEMENTS

Were it not for the *Springfield* (MO) *News-Leader*, and its editorial staff, this collection of columns would certainly not exist in this form. During my column-writing time there, two of my editors deserve special thanks. Early on, **Sony Hocklander** was a godsend. She moved the column from an occasional space-available status to a regular bi-weekly schedule. She was always cheerful and supportive, and made me feel like "part of the team." Thanks, Sony!

When Sony moved on, **Stephen Herzog** stepped right in without missing a beat. He continues to be efficient, approachable, upbeat, and a pleasure to work with and for. I've been lucky.

At *Oghma Creative Media*, **Casey Cowan's** vision for a different kind of publishing house seems to be succeeding beyond expectation, and I'm pleased to have become an "Oghmaniac!" From their corral of talented writers and staff, three of them were indispensable in moving the book from its infancy to adulthood. Upon acceptance, **Velda Brotherton** provided excellent guidance and advice in moving the original draft up the line. In the (first) final stage, my editor **Michael Frizzell** helped sprinkle his excellent advice and perspective into the mix with a perfect blend of acceptance, encouragement, and top-notch advice. Finally, when we agreed to add material and go with a longer book, **Jeremy Menefee**, our imprint coordinator, became my new editor in order to shepherd me through the rest of the process. The previous run of excellence never missed a beat! His excellence, perspective, and good humor provided the perfect final touches! If you're reading this, then it's a safe bet they all survived me, and for that I am pleased! Indeed, they have my gratitude and my respect, and, for each of them, I have doubled the amount I've left them in my will.

CONTENTS

—

Section 4 - Life and the Real World

—

Section 5 - The Magic (and the Magicians) of the Classroom

—

Section 6 - Beyond the Classroom: Administration and Support

—

Section 7 – Math is not a Four-Letter Word

Part A: Perspective

Part B: Thinking About Mathematical Thinking

—

Section 8 – Technology: Enabling or Crippling?

Section 9 – Politics & Political Issues

—

Section 10 – A Potpourri of Flavors

—

Foreword

Who knew? I've probably known Larry Campbell for close to three decades, but to me he was always a "math guy," a well-respected mathematician and mathematics educator who served as a faculty member at both the College of the Ozarks and Missouri State University in Springfield, Missouri. I had the opportunity to work with Larry on several projects in the field of mathematics education. Now, after reading *Spitballs from the Back Row,* I recognize that my mathematics educator friend and colleague is actually a latter-day Mark Twain!

What a perfect title for this book—*Spitballs from the Back Row.* Larry notes that he was perhaps too nerdy to ever actually craft and launch spitballs, but he has surely crafted, launched, and connected with the messages, I won't call them essays, provided here. These spitballs, and I actually think some of them come from the front row, are lessons about life. Consider just a few of the spitball messages that I just couldn't dodge:

- Kid Brothers and Big Picture Educational Issues... "Isn't it true that the best and most creative of solutions usually emerge NOT from one 'side' or the other, but rather from taking the best of both sides and seeking to move forward in a win/win fashion?"
- Hit Between the IZE... "While turning out educated

students is not a 12-year assembly line process (or is it?), it's worthwhile to ponder effects of education's natural tendency to centralize, standardize, and specialize on our schools. I have begun to call it 'getting hit between the IZE.'"

- A Tale of Two Classrooms… "Memorizing something is not necessarily the same as learning it."
- Learning about Learning… "We often hear the phrase 'live and learn.' I'm not sure it shouldn't be the other way around."
- Three Reasons Not to Hate Math… "Somehow, someway, much of this curiosity, creativity, and enjoyment tend to vanish by about 4th grade. Why is that? Is there any chance the 'system' teaches it all out of them?"
- Riding Bikes and Learning Math… "We MUST 'put them on the bicycle' earlier (brain teasers, puzzles, strategy games, etc.). AND, we must allow them to 'skin their knees' in the process."
- Ball Caps and Mathematics… "A final word to students — wear those caps however you feel you must. And don't just 'study your math' like you'd 'take your vitamins.' Ask your math teacher to let you in on the adventure. It can be as much fun as wearing your cap backwards."
- Epilogue… "It's a tall order. No matter how good the system (and it is aging, too), we are always trying to prepare students for the future, in classrooms of the present, with tools and perspectives that are frequently stuck in the past."

What a great read. Don't dodge these spitballs, these life lessons. They must find their way into the discussions that all families should have. And, such "families" include school and college/university faculty members, boards of education, and community organizations. Perhaps most importantly, find the time to read, digest, and discuss Larry's life lessons (L^3 for the math geeks). As Larry so eloquently notes in his message entitled *Making Progress in Neutral,* "our educational system demands so much from our teachers, especially in the way of time, that we are robbing them of some valuable and important time to just plain think and reflect, especially during the school day. And this robbery is not just hurting them, it's hurting their students in turn, and therefore, all of us in society." Expand this to our own busy lives—we all need time to make progress in neutral. The 30+ messages Larry has provided will force you to think, to discuss these education-related, and, yes, math-focused life lessons. Think of this book, *Spitballs from the Back Row*, as your opportunity to make progress in neutral. We all need such opportunities. Enjoy.

Francis (Skip) Fennell
L. Stanley Bowlsbey Professor of Education and Graduate and Professional Studies, Emeritus
Project Director, Elementary Mathematics Specialists and Teacher Leaders Project (EMS&TL)
McDaniel College, Westminster, MD
Past President, Association of Mathematics Teacher Educators
Past President, Research Council for Mathematics Learning
Past President, National Council of Teachers of Mathematics

Introduction

When I was a youngster, a "spitball" was the weapon of choice for the troublemaker in upper elementary school. You took a very small scrap of paper, chewed on it sufficiently, wadded it into a ball, and tried to surreptitiously throw it at one of your classmates.

Now, I confess, I was already too much of a "nerd" to enjoy the sport (I think I missed a lot in those days), but many of my buddies loved them. They were readily available, provided relief from bouts with boredom, and were mostly only a minor nuisance, even to the teacher. Unless, of course, an all-out "war" erupted. Now that was trouble for everyone.

I'm reminded of those days, as I collect these essays for this book. I've been working in and around higher education for over 35 years. Over that time, I've written educational columns in spurts, but I'm certainly not on the forefront of the national educational discussions. I am the equivalent of sitting in the back row.

This is not to complain. I enjoy chewing on these topics and throwing out these perspectives. Not entirely unlike spitballs from the back row, but hopefully not as annoying, at least in the same way.

On the other hand, I'm satisfied if these perspectives become minor "irritants"—not in the traditional negative sense, but rather in the forced-to-take-notice-and-think-about-it sense. Like a buzzing gnat. Or a spitball. This explains the title, and for that matter, tells a lot about the book.

Problems and issues cannot even begin to be "solved" if they aren't at least noticed. And there's so much in education that's happening by habit or reflex these days that the implications too often aren't even *noticed*, let alone thought about. Sometimes it takes a spitball to get one's attention. And therein lies part of the point. The selections here are not designed to convince. I don't claim to be *right*. They are designed to get attention, promote thought, and even (hopefully) discussion. So, read with pleasure and an open mind. Form your own opinions. Discussions—even debates—encouraged, but no arguments allowed.

As you can deduce, I'm enjoying lobbing spitballs out there to be thought about. I hope you'll enjoy reading them. It's fun, and it's relatively safe. Perhaps I'm making up for lost time as a 5th grader.

—

Before moving on, let's look a little at the book itself and its format. Here are some things you may want to know:

- Almost every selection in this book was a column in the *Springfield (MO) News-Leader* at one time or other over the past couple of years or more. Some of them have *also* been earlier columns and recent blogs, in other lives, but the versions in this book have been selected from that source.
- Naturally, I've edited, polished and even re-written some passages for the purposes of this book and its audience (e.g., most local references have been eliminated or clarified). But, I've deliberately chosen to keep the length and feel *roughly* the same as the original column. It took me a long time to learn that saying less is often saying more (a 600-word limit helps a lot with this), and I'd rather address a specific issue twice (as I sometimes do here) than get bogged down in babbling.
- Over the years, I've addressed overlapping collections of issues at several times, and in several ways. This cannot be avoided, as this is the way it is with educational issues. To classify

them often dilutes them. This has frequently made it hard to put a former column in just one of the given Sections. The various Sections are there for convenience, but they are rather artificial.

• All this will also make things occasionally feel a tiny bit familiar at times (often within a section, but also from section to section). I have not tried too hard to avoid this, any more than I try to avoid it in my column choices. Sometimes a hint of repetition is how things sink in, especially over the course of lots of columns. You may find your own viewpoints crystallizing not as a result of one column, but rather as a cumulative result of having read all of them.

• Finally, I hope you'll notice the chance the book affords for you to make further connections to other material—and with Oghma Creative Media—and for *us* to make connections with each other. It's virtually interactive! (Double-entendre intended.) These things happen in the three appendices (and elsewhere).

• Appendix 1—**Related Resources** is not only a summary of the various links that are mentioned in the selections, but also other Resources as well.

• Appendix 2—**Connections** provides a variety of ways that we can connect, including e-mails, my website(s), and more.

• Appendix 3—**Assorted Brain Teasers.** Lighten up! This is an *entirely optional* portion of the book. Ignore it completely if you want. But I think we've given just enough hints throughout the book of what kinds of fun might be lurking there that you might want to take a peek. On the one hand, you can do this in private, as no one will know if you look, and *no one* will grade you if you play with one or more at your leisure. See more details in Appendix 3, if you dare look.

Prologue

Years ago, my son made an interesting comment to me. He said something like, "you know, it's funny that 'education' is the one thing that everyone seems to support... and yet no one can agree on how to define it." He's right—and until that moment, I had never quite thought of it that way.

And therein lies the problem and the primary *raison d'etre* for this book. One of the hardest parts about discussing educational issues regularly is that while *everyone* is for a 'good education,' almost *no one* can define *what* it is or should be (go ahead, try it.). We're all using the same words with different images and concepts in our heads.

There is no clearly-articulated vision of what makes a 'good education,' except perhaps in each individual's (or group's) head, so there is little agreement on how to tackle some of the complex issues education faces—except, of course, in each individual's/group's head. As evidence, look at how politicians are in such disagreement on how to 'fix' our schools—and for that matter, *if* they even need fixing. And how those same politicians are quick to advance solutions to problems that exist, *at least partly*, in their own heads.

What is it we want our students to know? This question can be asked at the most general or the most specific level at any moment, any time. Do we want them to learn facts and procedures and titles and formulas and dates? Or do we want them to be able to learn how to do much more: to apply knowledge, to solve problems, to think critically, indeed to learn how to *learn*? Probably we want *all* those things.

Do we want them to learn 'values' in schools? And if so, whose values? Do we want them to learn to hear others' opinions and discuss both sides of an issue, or do we want to teach them what's 'right' (by someone's definition?) and have them never waiver in the 'correct' beliefs we give them? Do we want them to learn to be good citizens, and if so, what does that mean? Do we want to prepare them to make a living or to make a life (or of course, both)?

These are tough questions, and some of them are intentionally slanted in their wordings, for effect. But, together, they speak to an undeniable but-seldom-acknowledged fact: *education is a hugely complex process*. That process is incredibly difficult, no matter how we define and shape it, and issues that face educators today are almost never easy—or even well-defined. Indeed, they never have been.

Here we are, then. These, and more, are the kinds of questions I often ponder, and they are the types of questions and issues that surface in this book—sometimes repeatedly, and sometimes in different clothing and surroundings, as already mentioned. Sometimes they're addressed directly and sometimes they're hiding just behind that nearest tree.

And while I am certainly not afraid to share my own opinions to a point, the broader idea is to shine a brighter light on some of these issues, raise subtle and different perspectives, minimize misperceptions, and always try to provide balance. There are just too few folks listening to others' ideas, whether they agree or not. And when we quit listening, we all lose chances to make things better.

SECTION 1

What is Education, Anyway?

INTRODUCTION

We've probably all heard the old childhood riddle/joke: "How do you put 20 elephants into a Volkswagen?" The answer, other than 'why would you want to?', as you may recall is, "One elephant at a time!"

Stuffing elephants into a VW is roughly the equivalent of our task in confronting the problem question: "What is education?" The near-impossibility of the two tasks above is about the same. Really, the only way to approach it is "one elephant/issue at a time."

With education, however, the task is even harder. It's hard to get your hands on just "one issue" in education. They are all interrelated, almost by their very nature. It's like "Siamese elephants." We can't separate them, so how can we get them into the VW? But it is essential that we try. Even getting some partial success is exactly that—a (partial) success.

This first collection of columns, then, serves as a preview and overview of the remainder of the topics. It picks up on the general question in the prologue and tries to narrow it a little, to at least re-frame some things in smaller bites. The "Siamese elephant syndrome" will make some of this repetitive, as some "topics" will be subtly present in more than one column, in various leading and supporting roles, and in more than one light. But that is *precisely* the way it is when trying to deal with educational issues, so off we go.

WHAT DO WE WANT THEM TO KNOW?

—

At the root of everything we do in education are two important, fundamental questions:

1. *What* do we want our students to know?

2. *How* do we know when they've learned it?

On the one hand, these may seem obvious. On the other hand, they are often passed over lightly, *as if* they were obvious. This can be dangerous, considering the multitude of subtleties and sub questions lurking there.

Both questions above can be asked perfectly well (several times a day?) at any of the classroom, school, district, college, university, state, and national levels, and they permeate *all* we do at any of those levels. *All* the individual and collective answers are subjective, of course, and my/our answers might be different from yours/theirs. But if we aren't at least *consciously asking* these questions frequently, we are doing our students a disservice.

In this selection and the next, let's take a quick look at these two crucial questions separately. Let's take them in order, especially since it makes no sense to ask the latter question if we don't have a good handle on the first. We've touched on some of these areas before, but every perspective helps.

In the broadest of terms, the first question can be phrased (at any level) as follows: "As we proceed well into the 21st century, in a highly technological, instant information, social media society, *what is it that we want our students—future citizens—to **know?*** What should they be able to do well in this society, and how do we prepare them for that?

In my academic discipline, for instance, this might be further translated: do we want our students to learn *skills* (times-tables and long division with paper/pencil, algebraic manipulation, etc.) or do we want them to have the *power* to formulate, tackle, and solve real world problems using appropriate technology? Obviously, it shouldn't be an either/or question: we want both, of course. But how much of each, and which comes first? (Do we learn to build things after we learn to hammer/saw, or do we pick up those skills as we learn to build things?)

In other disciplines and areas, the translations are similar, but the question itself continues to be deviously difficult to answer, even for "experts." As we have mentioned and will mention consistently, the "answers" are not easy, *and* they keep changing with the times.

As always, there's another fly-in-the-ointment here. As we all help prepare our young students to be educated, well-rounded adults, and good citizens, the fundamental question "what do we want them to know and be able to do" extends far beyond the more "academic" (and relatively safer?) areas. How, for example, do we want our students to form their own opinions on important subjective areas (politics and religion, e.g.?), and how do we want them to discuss (and defend?) them reasonably (and with civility) with others?

If, as a society, we are going to extend the question beyond academics (as I think we probably should), then we must also ask: to what extent can "education" help in those areas, and to what extent should those questions be "hands off," and left to parents and society in general? That's a delicate, controversial question, isn't it? But, regardless of your own feelings, note how any of our "answers" speak directly to what we think it means to be "educated."

We'll look more at the second question (*When Do They Know It?*) in the next selection. But for now, it's worth considering, re-considering, and discussing: what do we want our students to know?

* *For another look at "what should our students be learning?" see* ***Education in an Information Age*** *in Section 3 and* ***Where Do We Go From Here?*** *in Section 8.*

...AND HOW DO WE KNOW THEY KNOW IT?

How *do* we know when students have learned what we think they should know? This question follows the question of the last column. "*What* do we want students to know?" We'll visit, more than once, about how difficult this question is to answer, as well as the interesting fact that, almost without exception, the more meaningful the topic or lesson to be learned, the harder it is to assess. We'll even note that often the reverse is true: The easier something is to assess, the less meaningful it usually is.*

It is crucial to understand that evaluating and assessing learning is so much more complex than it is often given credit for, especially by politicians. This partially explains the difficulties of various legislative initiatives of the past few years, no matter which side of the aisle advanced them. Easy answers aren't necessarily meaningful *or* helpful.

To illustrate some of this, I've adapted a short True/False quiz about learning and assessment.** You won't be "graded/assessed" on your answers. (Did you note your reaction to the word "quiz" just then? It's significant and relevant, but must wait for another discussion.) Indeed, there are no "right answers" to this quiz—only opinions, but as you decide your own answers, notice how your answers automatically influence your views of assessment, and even education itself.

1. Students learn only by imitation and memorization.
2. First we teach, then we test.
3. Learning a subject means mastering a fixed set of skills; therefore, tests should focus on whether students have mastered those skills.
4. Objective, multiple choice tests are the best and most reliable instruments to measure learning.
5. The purpose of assessment is to determine which students "have it" and which do not, and then to assign grades accordingly.

6. Alternative forms of assessment are less objective than traditional forms and are simply the latest fad in ducking educational accountability.

Interesting exercise? Did you feel yourself pulled in opposite directions? Did you want to answer a question "false" in general, but realize the assessment can of worms you opened if you did so? It's the problem educators face all the time, of course. Obviously, any good assessment plan or scheme, whether a teacher's, a school system's, a state's, or a nation's, will have to have grappled with tough issues like these as—and even before—it begins to tackle assessment and accountability.

It is worth repeating that there are no right answers to these questions, but there are certainly different opinions. And those opinions affect and define what we call education. For example, each "TRUE" answer above represents a relatively traditional view of education—and for that matter, a relatively "safe" view of assessment. Each "FALSE" answer, on the other hand, makes the job of accurate assessment that much more difficult. One approach is easier, the other harder. The key question, though, should be: Which approach has the best chance of helping us make meaningful evaluations in the long run?

How many times have we encountered this fact: there are no easy answers in education, and often no "right" ones. It becomes that much more important that we're at least *asking* the right questions. If not, we're not serving our students, regardless of answers. And, in assessment, as elsewhere, even determining the "right questions can be a tough and thankless job.

For a deeper discussion into this phenomenon, see* ***Meaningful or Measurable*, *in Section 3.*

*** *In preparing this quiz, I've used a reference called* "Mathematics Assessment: Myths, Models, Good Suggestions . . ." *published by the NCTM, but I have edited out references to math.*

KID BROTHERS AND BIG PICTURE EDUCATIONAL ISSUES

—

One of my brothers rarely agrees with me on political, theological—or educational—issues. And he's not afraid to share his takes on those subjects, which is good. This makes it both worthwhile and fun. We have both (finally) learned that our differences yield valuable discussions, as we each hear points of view that run counter to our own.

Recently he and I were discussing a couple of my education columns. He said *something* like, "You have some good points, and you're balanced, but you never come down on one side of the issue. You need to give us your own take." That's not verbatim, but I don't think he'd disagree with the gist of it.

I couldn't help but grin. He's right about the "not coming down," of course. I plead guilty. But I'm not sure I agree with his conclusion that I *should*. It later dawned on me, in a marvelously ironic way, that when it comes to educational (and often other) issues, the only place I *do* 'come down hard' is on the side of the firm opinion that you *can't* come down hard on big educational issues! The bigger the issue, the harder it is to take an either/or position.

When it comes to tackling big issues, and solving big problems in society, isn't it true that the best and most creative of solutions usually emerge *not* from one 'side' or the other, but rather from taking the best of both sides and seeking to move forward in a win/win fashion? The point is, those who are tasked with solving these problems cannot do this until/unless they *understand* the truth that is almost always on each side. When it comes to big issue problem solving, it becomes a time not for "having takes," but for asking questions. It is a time not for "coming down," but for "coming along." It is not a time for convincing, as much as it is a time for listening.

Perhaps you're asking (along with my brother?) what this has to do with education? It has *much* to do with education, and in two very big ways—one of them "obvious" and the other much subtler.

First, it is almost always the case that profound educational issues—ones that really matter—are of exactly the type described above. One can't begin to make authentic progress on them until one *understands*, and then *honors* the two (or more) "sides" that make something an issue in the first place. For my own part, I prefer to openly spotlight both "sides" of some issues I've encountered. This has been and will be viewed (as with my brother), as "dodging the issue," but I view it as contributing to more meaningful grounds for discussions. More irony: perhaps both those perspectives are "right!"

Let's not forget the subtler question: what are we teaching our young students—our future citizens—about solving big problems, about seeking—and hearing—input, and about moving forward together to seek win/win solutions? Some would say "*How* can, or even *should*, we be doing that?" And (surprise), I grant there's no easy answer. There never is. I guess I would also notice that, whether *we* address it or not in the classrooms, students are *learning* from watching us. I sometimes fear for what they're learning. Shouldn't we start giving them enlightened guidance now? Do we want to just hope they're lucky enough to have a brother that disagrees with them?

HIT BETWEEN THE "IZE"

—

The remark came, almost in passing, in a chamber of commerce leadership session I was chairing years ago, but it has stuck with me. The speaker was listing various trends and how they affected our area of the Midwest. He noted that not long after the turn of the 20th century, industry was learning the value of specializing, centralizing, and standardizing in its factories. He claimed that education had soon begun to follow the same pattern.

Interesting. One can certainly see the truth there, and while turning out educated students is not a 12- year assembly line process (or is it?), it's worthwhile to ponder the effects of education's natural tendency to centralize, standardize, and specialize on our schools. I have begun to call it "getting hit between the IZE." (Yes, I know that "getting hit *among* the IZE" is grammatically correct in this case, but it tends to ruin the play on words, so humor me, okay?)

Certainly, getting hit between the IZE is not all bad. Like factories, centralizing schools and their operations is much more cost-efficient. Standardizing operations, curricula, and tests, both locally and nationally is much more uniform, and theoretically contributes to a certain degree of improved accountability. And, specializing in subject matter should, in theory, produce better workers (teachers), with more/better knowledge in their various content areas. And none of those things is bad.

As noted, though, a school is not a factory, and we are not turning out uniform products on an assembly line—nor would we want to. So, there are also drawbacks, as well.

We are dealing with individual students, thus, the bigger (and more centralized) a school system gets, the easier it is for students to get lost, to fall between the cracks, to lose "individual" treatment.

As a natural result, standardization begins to kick in. Rules invariably get made and applied uniformly, regardless of special cases. That can make things fairer on the surface (justice, after all, is blind), but, as in any situation, it takes away the ability to handle special cases, and the bigger any group gets, the more certain that there *will* be legitimate special needs and cases.

Standardization, for all its potential can also be unexpectedly cruel. Allegedly (though some scholars doubt this) Thomas Jefferson once said, "There is nothing more unequal than equal treatment of unequal people," and certainly not all students are the same. Students anywhere may take equal standardized tests, but can all those students possibly have similar lifestyles, backgrounds, and cultures? Will that one test tell who is better educated and ready for the real world? (Will *any* test do that, I wonder? But that's yet another set of thoughts.)

Finally, the drawbacks of specialization can be very subtle and less noticeable. The more fragmented a student's "education" gets, the more he/she tends to see the parts, rather than the whole. In my area of mathematics, we give kids a year of something called "algebra," a year of something called "geometry," and so on. Rarely do we show them how they all fit together as tools. No wonder so many kids have trouble with (what they view to be) "math" in general.* This is not only true within subjects, it applies to viewing *all* "subjects" as part of an overall process. (It's worse in elementary grades, but...)

We've come a long way since the days of the one-room school, and I'm not suggesting we should return. On the other hand, if the advantages of size and similarity deter our ability to effectively educate individuals, then they cease to be advantages. At that point, getting hit between the IZE in our schools becomes fatal for all of us.

* *For more perspectives on the area of mathematics (and how we teach it), see Section 7.*

ON THE OTHER HAND...

I once ran for School Board in our town. The good news is that I made a fairly strong showing—I finished right behind two very popular incumbents. The bad news, which I later decided was also good, is that there were only two slots open.

I mention this to say that one of the constructive critiques I received of my candidacy was that I was too "wishy-washy," too "on the fence," as it was called, in various issues. (Does this sound like my brother in the second essay in this section?) I confess this critique was valid, but I'm still not sure it was a weakness (I can be good at rationalization). I have always maintained that issues encountered in education are *much* more complicated than they often appear (especially to Congress), and solutions to problems are not always straightforward. Indeed, they can be fraught with hidden dangers and unintended consequences. It is important to be "on the fence" for a while at times—the view and the perspective from there are often better. Balance is often needed as part of a win/win solution process.

I'm going to reluctantly demonstrate this by arguing—or at least mentioning—a different side of an issue I discussed in the previous selection. There, I made a case for the subtle dangers involved in standardizing our curricula and testing in educational processes. But the other side should be given fair treatment as well, for this issue is not easy.

Since the rapid expansion of our country during our frontier days, we have become a strong "local control" nation, and we are fiercely proud of it. No one knows better, so the argument goes, what is best for *our* kids than our local school boards. And, of course, there has always been truth to that argument. But, the argument has limitations.

For years, I taught at a small liberal arts college whose associated public-school district was in a small town that still prides itself on its small—even rural—roots and values, and works to preserve them. This can be refreshing! But as a result, it was not unusual for the make-up of that school board to consist of three faculty members from the college—often with terminal degrees—and three local graduates who may or may not have gone *any* further in their schooling than high school. This is neither good nor bad but imagine the diverse perspectives about "what is best" that were inherent in the backgrounds and viewpoints of those board members. Diversity is good of course but can make it difficult to reach consensus.*

Our nation's schools have, for some time, taken considerable heat over our relatively poor showings in international comparisons of various test results. It is worth noting that, almost without exception, countries that "beat" us are countries that have standardized (if not nationalized) curricula. So, it seems to be a case of "pick your poison."

In passing, I'll note that my own discipline of mathematics education—through the National Council of Teachers of Mathematics (NCTM)—has addressed these contradictory concerns in forward-thinking fashion. The NCTM was the pioneer of the "Standards" movement, in which rigorous "Standards" of what students *can* be learning and achieving are presented as a vision, without trying to dictate how these goals are achieved or even if they should be adopted.

The point in all of this is *not* to argue for or against standardization in general. Obviously there are pros and cons either direction. Instead the point is to re-emphasize that issues in education are not easily solved, and should not be treated as if they are.

* *Update: On the other hand, this same community recently elected its newest board members in such a way that* all six *are from the same church. (Rumors of a hidden agenda abound. Imagine.) Will the reverse now be true? Easy-to-reach consensus, but much narrower perspective?*

THE SHELL GAME

The Greek philosopher Heraclitus is credited with saying, "Much learning does not teach a man to have intelligence." As if to illustrate, I once received an anecdote from a frequent reader. The scientist Isaac Asimov tells this story on himself. His car mechanic runs a riddle by him and catches him on an obvious blunder. The mechanic claims he knew he'd catch Asimov. When asked why, he replies, "Because you're so stinkin' educated, doc, I knew you couldn't be very smart."

We've all heard similar stories or quotes, I suspect, but this difference between knowledge and intelligence, or wisdom, is worth exploring in a slightly different way, as it relates to learning and our educational classrooms.

In the Mar/Apr '89 issue of *Missouri Schools*, there appeared an article titled "The Shell Game." I still have some quotes, as well as the column I wrote for our local paper back then, so I think I'm safe to summarize.

The author's main premise seems to be that teachers should inform their students of precisely what material should be "mastered" before a test, and then test them on exactly that material. Any other approach, he claims, makes the teacher guilty of what he calls "the shell game," forcing the student to try to choose what they are *supposed to know* from too many pieces of information.

He accused some teachers of engaging in "unscrupulous classroom games" when they play this shell game and he urged administrators to correct this practice and to "question the teacher's preparation and planning" and demand to know the "dates, facts, procedures... emphasized during the lesson."

Gracious. Where do we start? I certainly don't believe in trying to trick students, but don't two questions immediately scream at us?

1. **"Dates, facts, procedures?" Is that what they are "supposed to know?"** Do we want them to know when,

where, and even who invented the light bulb (author's example) or, for example, would we like students to learn and profit from Edison's avenues to success (even the unfruitful ones), and perhaps even explain why/how the light bulb works? (Okay, we *might* want both, but if we could only choose one, which would it be?) And, in passing, if they learn the latter, don't they usually pick up the former as icing on the cake?

2. **If we test them on "facts" we've just made sure they "need to know" what are we really assessing?** Are we assessing their "knowledge" or their ability to memorize for the short term? For my money, one of the reasons the whole assessment question is so controversial is that it's just not as easy as the author paints it, and to believe so is to wade into a heap of trouble.

Perhaps the real shell game occurs any time any of us—teacher, administrator, parent, or citizen—hide from ourselves the fact that learning is more than collecting facts, that teaching is more than doling out facts, that assessment is more than testing over those facts, and that being educated is more than knowing a collection of facts.

For more thoughts on collection of facts, see ***Education in an Information Age*** *in Section 3 and* ***Where Do We Go From Here?*** *in Section 8.*

BEYOND THE SHELL GAME

Recently I ran into a quote on a t-shirt that makes a perfect segue for the previous "Shell Game" thoughts about collecting facts, right answers, assessment, and education in general. (It also makes a great preview for some of our later thoughts about calculators and technology in all fields, especially in Section 7.) The quote *happens* to be about mathematics, but don't run away. We'll quickly generalize to all fields of education.

The quote goes like this, "Good mathematics is not about how many answers you know—it's about how you behave when you don't know the answers."

This is more than clever wording. Think about it. Would we rather have a student in algebra who can eventually factor every polynomial (right answers) or one who can recognize when algebraic concepts might be helpful in tackling (and maybe solving) a real-world problem involving math? Would you rather have a student who can take a square root by paper/pencil, or would you rather have one who knows when a square root is called for in the solving of a problem?

As mentioned, we need not restrict this idea to mathematics. Would we rather have a history student who knows that Abraham Lincoln was assassinated by John Wilkes Booth (who) at Ford Theatre (where) in April 1865 (when), or a student who might have forgotten one or two of those "facts," but could thoughtfully think and write about the effect of the assassination on our country after the Civil War?

Numerous examples abound. You get the idea. The point is that becoming educated is *not* (only) about collecting "right answers," which is to say information. Especially in our "information age" when these facts are readily available by computer or smart phone. It is more about how we use the information we learn. Are we satisfied only with "right answers" or do we want our students to learn to think as well? (For more thoughts along this line, see ***Education in An Information Age*** in Section 3, e.g.)

Obviously, it's not either/or. We want the ideal blend of *both.* We want students to learn some *basic* information, *and* to learn to use it effectively when they need it. But there are two traps here that can be *very* subtle. First, there's always the question of what's "basic" (often controversial, by the way). And then there's this very subtle by-product. *More is not always better.*

It is not unusual for school systems, colleges, and/or state school boards to require *more* courses from students. More years of math, more years of English, more specific courses, etc. This is usually done in the name of "high standards," which is fine. But, if not considered carefully, it can also be counter-productive.

In the first place, "more" is not always needed for *all* students. I'm about to commit possible heresy in my own discipline, but it simply is not the case that *everyone* needs three or four years of math in high school. Another classic example here is the university that "requires" college algebra of *all* students to graduate, rather than a more helpful "Math in The World"-type survey course for some less-math-oriented career paths.

Second, requiring more "stuff" of all students can severely limit students' chances to try other areas of enquiry, and to discover a passion or interest they didn't realize they had. Limiting these opportunities at this crucial juncture in their lives can be damaging not only to the students, but to society as a whole.

As always, a final word of perspective: this column is not about "math" or "history" or "course requirements" specifically. Instead, it is yet another example of asking ourselves continually *what* we think education is, and what that means for our classrooms.

MOVING THROUGH THE SYSTEM

The story goes that if you say "good morning" to a class of elementary students, they will return the greeting, if you say it to high schoolers, they will respond "what's good about it?" and if you say it to a class of college students, they will write it down.

I've had *some* experiences with enough students of all ages to recognize the kernels of truth prompting that humor. Naturally, it's risky to draw general conclusions too rapidly, but it's worth exploring this a little further.

Years ago, I/we ran some "Fun with Math" classes for grade-schoolers on Saturday mornings. As you can imagine, these kids had vast storehouses of energy, and not surprisingly, it was hard for them to sit still. (Once again, I was reminded of our system's criminal tendency to load down an elementary teacher with dozens of these kids in each class.) Ah, but this energy and enthusiasm translated into a genuine curiosity that was fun to harness, an eagerness and willingness to try new things and very little of what we would call "fear of failure" which would inhibit their learning. As a result, we had a great time and learned a lot of neat things (though I needed a nap when we finished!). (For more on some thoughts about this experience, see *Pondering a Class Size of "Only" 20* in Section 6.)

On the other extreme, at least age-wise, are the adults I often encountered who were back in school to finish a degree (or certification), often under trying circumstances. Over the years, I really came to love working with these students. Clearly the most "dedicated" of the various ages, they were focused, knew what they wanted, and were often willing to overcome their fear of "math" in order to get it. They also exhibited curiosity, but in a subtle, different way. They wanted to know how "this stuff" applied to their world, and what they were "supposed" to know for the test. They were dedicated, to be sure, but alas, the pure fun of learning had long since gone. They were the most likely of the age groups to view education as "only" a rigorous (and often costly) path to an end. Ironically, these were often future teachers.

Between my sojourns at different institutions of higher education, I taught a year of high school Senior Math—as a service, and as a fun experiment. Interestingly, indeed tragically, these young adolescents (and their college counterparts I've dealt with) were the least "alive" of *all* the age groups. Here were the experienced veterans of over a decade in "the system," and they often moved through their classes as if responding to a distant unspoken set of rules. Curiosity about the subject hardly ever surfaced, not because it was never present, but because the students had learned that curiosity doesn't pay, isn't worth the energy, and doesn't get "the grade." With some effort, I could occasionally see—even induce—the sparks of fire that used to blaze back when *they* were youngsters learning to sit still, but it was an uphill battle. They were usually great kids—and bright, too—and I enjoyed them. But to most of them, getting educated meant riding out the system mechanically for their remaining years.

Yes, I know, these are over-generalized snapshots. At the same time, I've seen the same distinctions and differences often enough to know there's too much truth in this evolutionary overview of our *system*. What happens to our kids as our system *educates* them? And *what* does it teach them? Do they have to lose their eagerness and curiosity when they learn to sit still? (Do they *really* have to learn to sit still?) How can we nourish, rather than stifle, their love of learning?* As always, these are tough questions, and answers are buried deep. But shouldn't the search start with the next "good morning?"

* *For a refreshing view of the "love of learning," see* ***Live and Learn...*** *in Section 2.*

SECTION 2
Lessons Along the Way

INTRODUCTION

The best lessons, of course, rarely occur in the classroom. Put another way: The best lessons usually occur in the classroom of life. Sometimes these experiences affect you right away, and sometimes, it's a while before you realize the value of what happened. If you're lucky, you look back and realize and learn.

Some of my most meaningful lessons about education happened to me when I was least expecting them—right in midstream, so to speak. Some of them are worth sharing, partly because of their impact on me and my career, but also because, if you're aware in these situations, you see an issue exposed, presented, or put in a new light in ways that no academic discussion can do. That's what they did for me.

In this section, we explore some things that affected my view of various educational issues, sometimes immediately, sometimes later, and usually for the duration of my career. They are arranged in *roughly* chronological order; some go back to my own school days, and then progress through stages of my career.

In life, as in the classroom, sometimes hands-on experience is the best teacher.

SEARCHING FOR RIGHT ANSWERS

This occurred decades ago, you understand, so various details are fuzzy, like watching a movie with the wrong pair of glasses. Still, I remember the basic experience, and some of it has been with me as an unsolved riddle since then. I don't claim to have a "right answer."

I was in high school—perhaps a junior? As it happens, this was a math class, and it was taught by my favorite teacher. We had just finished a 5-question exam, one of those tests where each question built on the information discovered in the previous problem. I left class feeling confident.

Imagine my surprise the next day when the test was returned with a D grade. I couldn't fathom what was happening. Surely there was a mistake?

There was a mistake, all right. Further examination showed that I had made an arithmetic error in the very first problem which affected the answer to the first question. Because of that, all the remaining questions on the test had "the wrong answer," though it turned out I had worked each of those problems in the correct manner.

I was beset by a whole a cauldron of emotions. There was wounded pride (I was too grade conscious, even then), there was anger (but at whom, and why?), and these began to morph into a huge sense of injustice.

Cradling all these emotions, I cautiously approached the teacher afterward. This was my favorite teacher after all, and I was too nerdy to stir the pot too much. Besides, I wasn't sure I had a case anyway. Perhaps I was merely looking for sympathy?

I tried logic: A) I had worked all the problems "correctly," and the mistake was a "slippery one," as she called them herself. B) The grade was influenced by the timing of the error—had the arithmetic goof occurred in Question 5 rather than Question 1, the test grade itself would have been much better, though the situation was nearly identical.

She replied, of course, with time-tested logic of her own. It went something like, "I understand Larry, but in real life, it doesn't matter where an error occurs if it throws off the final result. You don't get 'partial credit' if your building falls down." I couldn't (and still can't) argue *that*, of course. I admit that's certainly true, then and now.

It goes without saying the grade stood. Still, I know she was sympathetic, and I think she secretly agreed with parts of my case. (In language that I couldn't have articulated then, I think she was struggling with the "does-the-assessment-match-the-learning?" question herself, but I'll never know.)

As an educator, I never fully decided which one of us was "right" back then. Indeed, I've come to believe that the devil here is in the dichotomy. I think we were both "right," and we were both "wrong." Interesting paradox.

This influenced me throughout my career in two ways. First, how do I prepare students to "build solid buildings," yet look behind temporary right/wrong "answers" to assess if they're authentically learning how to do that? (This question is universal, regardless of subject matter.)

And, stepping back a little, this incident was a microcosm of bigger issues. In my own memory, we were both right, we were both wrong. Isn't that the case with so many issues/dilemmas in education? Isn't there always "right" on both sides? (If there weren't, would it even be an "issue" in the first place?)

Perhaps our searches for the "right answers" to educational dilemmas should be replaced with broader searches for win/win solutions.

For a follow-up to this column, see ***Back to the Future*** *in Section 4.*

A TALE OF TWO CLASSROOMS

I was in the first year of my master's program and taking two math courses from one of my then-favorite professors. That semester he chose to try a new experiment: he decided that, in both these classes, he would give *unannounced* hour-long exams!

Not just pop quizzes, mind you, which we may all have encountered, but hour exams, the collection of which comprised a sizeable percentage of our class grade. He said he had randomly picked the dates of the 3 exams in each class ahead of time, and it didn't matter when they fell. If they ended up weeks apart, or on two successive class periods, so be it.

It was a radical move. Take a minute to imagine the effect on the entire class environment for the semester. As a humorous side-effect, I don't think there were any attendance problems in those classes that semester.

Can you also imagine the ramifications on a student's study habits in the class? You couldn't really "cram" for every class meeting. You *had* to "keep up" and you *had* to learn the material.

I don't recall too much about that semester (hey, it was a while ago), but I remember this: I learned as much or more in those two classes as I did in almost *any* other math class in my long sojourn of seeking degrees. Further, I also learned a lot about what it means to *learn* a topic in general, and how one does (and *doesn't* do) it.

I think that experience, as much as anything else, may have started my career-long pondering on the nature of real *learning* in our educational system.*

Compare this experience with another one, two or three years earlier in my undergraduate days. One fall semester, we had a brand new professor fresh out of grad school teaching one of our upper-level classes. He was brilliant.

As is stereotypical, however, he had not (yet) developed the ability to share his knowledge well with those of us not yet on his level. I remember all of us frantically trying to keep up as his thoughts spilled from his head to the blackboard, and then rebounding to our notes.

I remember memorizing the proof of one theorem, and later congratulating myself for reproducing it flawlessly on a test. But, of course, I had *no idea* what I had written, nor what the theorem really meant.

That may have been the semester in which I learned the *least* of any math class in my career. I didn't really learn to appreciate the material in that class until I began to teach it years later, when, with some experience, the subject's power—and beauty—really hit me.

There are a dozen implications we could pursue here, and I suspect you've thought of many of them. For this column, however, let's start with one obvious one: *Memorizing* something is not necessarily the same as *learning* it. Think about that, for there is a surprising result that must follow, and it is absolutely crucial: "testing" (*especially* testing "facts") is not necessarily the same thing as "evaluating learning."

Perhaps we take it for granted that student *learning* is at the heart of everything educational, but *learning* is simply not a forgone conclusion. Authentic learning is not always easy to achieve, and almost never easy to evaluate. Pondering this fact and creating environments in which learning can be achieved and then evaluated is one of the ongoing challenges of any teacher or institution.

For a peek at another experience that influenced my views on learning, see the next selection,* ***Learning about Learning.*

LEARNING ABOUT LEARNING

—

It was September or October of 19xx—I'd rather not be precise—and I was in my first full semester of teaching out of graduate school. I had been assigned a Differential Equations class, and quite honestly, I was still a little unsure of myself. I had not really had any differential equations since my own undergraduate course, and back then it had been my "math class from hell," for several reasons. I'll confess to some A's in my 4 years of undergraduate math, but Diff-EQ was *not* one of them.

As you can imagine, it felt like I was keeping roughly twenty minutes ahead of my students, and I was struggling with many of the homework problems I felt I "should" assign. I was a new professor, I was rusty on this topic (at best), and I was frustrated and panicked, just like a student. Indeed, if I'd been a student, I'd have been lucky to be earning a B.

One afternoon, out of the clear blue, I had a revelation which was to me earth-shaking, but I'm not sure I can capture it fully in print. "Wait a minute, Campbell," I heard myself saying, "You have a doctorate in mathematics. That doesn't necessarily mean you are automatically good at differential equations, but it *should* mean you have the confidence to relearn this material on your own, grind through any of these homework problems, and then help students learn the material."

For some reason, this change in perspective, this shot of confidence, was *exactly* what I needed. From that point on, I was fine, even confident, teaching the course. Truth be told, in fact, I think I did a pretty decent job after that.

Not only that, the change of perspective carried over into a lot of my later courses, and certainly into my years of working with future teachers, as we talked about their own learning (especially in a subject they weren't always crazy about), *and* their future teaching of youngsters.

The lessons from this story aren't really about my newness as a professor or my rustiness in one subject. They are more about the nature of learning (and therefore teaching) itself.

It simply is true that learning—in academics or in life—*can* be a struggle. Sure, it *can* be easy, and it *can* be fun, especially if you're excited about and interested in the "topic" at hand. But if you're rusty or unconfident or feel you're being forced (by outside circumstances) to learn something, it can be a tough and rocky path.

If you're a teenager wanting to learn to drive a car, it can be easy and fun (sometimes too much so!). But if you're an older adult wanting to learn to play the saxophone or ride a unicycle, it may not be a quick process, regardless of desire. Trust me on this, for I have tried—or am trying—both of those!

Awareness of this fact can make all the difference in perspective and attitude, and therefore in the potential learning itself.

And as teachers, in academics and in life, this has striking implications. As teachers, we need to be aware that students will learn our subject(s) at different speeds and motivation levels. And allowing and *honoring* their lack of confidence can actually help the learning process in the long run. To succeed with a student when learning isn't easy is one of a teacher's greatest satisfactions.

So, here's to learning, and here's to teaching. Both can be complicated, but both can make our lives more enriched.*

For still more on (lifelong) learning, see* ***Live and Learn* *later in this section.*

MEANINGFUL OR MEASURABLE?

—

Our goal is to make the meaningful measurable, not the measurable meaningful.
Shirley Hill, Past President,
National Council of Teachers of Mathematics, circa 1991

It was late 70's and I couldn't have been more than two or three years into my first position. The entire faculty was at a late-summer Faculty Retreat at a spot removed from our campus. Our assigned exercise that morning was a good one: we were to come up with a List of the Top Ten Things we wanted of and for our graduates after they graduated.

Naturally, the task generated excellent discussion, and we managed to whittle our list down to ten (with several differing opinions, of course—as there should be).

The list included wonderful attributes *like* [these are *not* verbatim] "ability to think critically," "ability to effectively articulate points of view while listening to and honoring others' views," "ability to be good, informed citizens, while exploring and understanding issues," "willingness to serve society and our neighbors," and on and on. It was an impressive list.

It was then that the *insight* hit us with the force of a moving train. In looking at our list of Top Ten Attributes, it became clear to us that *none* of them could be precisely evaluated! How could we reliably measure that our graduates became "good citizens" or "critical thinkers," or "good communicators," or...?

I'm going to deliberately go *way* out on a limb here: Is it possible that, in an educational setting, almost *nothing* that's valuable (to a discipline or to society) can be measured effectively? And conversely, is it possible that almost *nothing* that can be evaluated with paper/pencil or standardized tests is actually valuable in the long run?

The implication here may be over-stepping even my own beliefs, but perhaps we need the extreme position to jar us into looking realistically and thoughtfully at the situation.

This perspective has *huge* implications for education in general, *especially* as it relates to the way Washington attempts to handle education "reform" (and both sides of the aisle are guilty).

Personally, I consider this insight—to the extent that it's true—to be good news. I'm a progressive, but let's face it: American education has lived with this particular brand of "fuzziness" ever since the time of Jefferson. It's only been relatively recently that we've been hit with standardization, specialization, nation- or world-wide test results, teacher "accountability," etc. Clearly, *none* of these things is bad in/of itself, all of them can be helpful, and all are as well-intentioned as they can be. But in the aggregate, they *may* be producing the long-range opposite of their intentions. And we must recognize it.

I won't advance "fix it" opinions here and now—another time, perhaps. There are no "silver bullets." But, as a society that values education, we must all take a huge perspective leap outward and ask ourselves some tough questions. Is it worth sacrificing meaningful to get measurable? (Or vice versa!) Where is the balance between the two? And what kinds of non-traditional creativity will it take to achieve it?

LEARNING BY TEACHING–AND TALKING

I was in the very middle of one of my first Calculus I classes that I ever taught, and we were going over a concept that sometimes causes trouble. I was going rather slowly, trying to carefully navigate the early rough waters with them and make sure they understood. I was in the very middle of what was (from my perspective) a particularly enlightening explanation, when the weirdest thing happened. It was *almost* as if time slowed, my mind stepped back, and a light bulb came on: "So *that's* why that works!" For one brief second, I was so incredibly delighted that I almost burst out laughing for joy at my newly-found deeper understanding.

I *think* I disguised all this to the class without missing a beat, pulled myself together internally (somewhat regrettably), and proceeded to finish the discussion/explanation and move on with the class.

That was, I think, my first confirmation of the almost paradoxical fact that we learn something better when we teach it. As the Roman philosopher Seneca said, "While we teach, we learn." I suspect we all have experienced this phenomenon in our own diverse ways at various times.

But I want to expand this a little, as the concept is broader and even more useful when it comes to learning, especially in the classroom. We may indeed learn a familiar concept *better* by helping someone else learn it, but we also learn a brand-new concept better when we *discuss* it.

The term used in educational jargon (and elsewhere) is *discourse* and it is immensely valuable in the classroom as students grapple with brand new concepts. Rather than simply listening to a lecture from a stand-and-deliver presentation, students are often helped by being able to talk about a concept. I know from experience that this is especially true with concepts in mathematics.

This can happen in a whole-class discussion, but it is even more helpful in student small groups. There, at least three things can happen:

1. Students are often willing to ask questions they wouldn't ask the teacher, and then get answers.
2. The very act of trying to articulate something often makes it easier to understand in the long run.
3. Students can, as a group, identify what is still fuzzy to them, and ask the teacher later.

Teachers don't use this technique as often as they/we perhaps should, because we often fall prey to the enticing belief that we can "cover" more material if we do all the talking. But "covering" the material isn't always the same as students *learning* it, and discourse often helps that happen better. As one of my colleagues once said, "I quit worrying about what I was covering and starting worrying about what my students were covering." (See ***Covering Material*** in Section 5.)

LIVE AND LEARN... OR LEARN AND LIVE?

How can a brief look from one eighth-grader be so memorable and so instructive?

I was just beginning a year of a non-traditional sabbatical spent working with teachers, students, and even parents on problem solving. I encountered the young lady in question in the very first period of the very first school I had visited that Fall Semester.

She had been particularly "grabbed" by one of the fun problems we'd worked on—the "calendar cube problem*," we called it—but hadn't gotten a solution before it was time for her class to move on to the next period. This had not bothered her in the least... in fact, she had not even wanted to hear the hint the rest of her small group had requested a few minutes before.

For some reason, this girl's unique brand of subtle, unconcerned confidence as she left the workshop stuck with me all morning. When I happened to run into her (almost literally) in the crowded craziness of a junior high cafeteria at lunch, I asked if she had figured out the problem's solution.

It was then I received the memorable look. Her chin fell almost to the floor, and the sadness was evident in her voice. "Someone *told* me the answer," she replied plaintively.

I was speechless. I think I just stood there as she walked off. Here was this perfectly marvelous eighth grader for whom the independent solving of the problem was more important than the actual answer itself!

It would not be the only time that year that I experienced such a mini-enlightenment, nor would it be the only time I would learn much from the students and teachers with whom I worked.

In the decades that followed that workshop, I've told that story innumerable times, especially in workshops and classes with current and future teachers, as we discussed the type of environment we'd like to encourage as we work with students.

But today, I'd like to focus on the wider implications for all of us, no matter how long it's been since we were in school.

How long has it been since any of us really experienced the excitement of learning something new? Not something we *had* to learn, mind you, or something we would be tested over, but something for which the knowledge gained was valuable simply because we wanted to know it and/or because the actual acquiring of the knowledge was part (maybe all?) of the fun? Probably too long for most of us. Interesting food for thought, isn't it?

I still occasionally think of that eighth-grader-who-is-now-over-forty. Is she still curious and confident about the little problems she faces? Is she still excited about learning in general?

For that girl, at that time, the classroom still held much of its magic. How long did that last? Soon, for her, as for all of us, life undoubtedly became more of a classroom than the classroom ever was. Does it still hold the magic? Does she—do we—still experience the joy of learning?

Here's a resolution for all of us, New Year's or not: Why not resolve to learn something—or several things—new and exciting this year? Something that really grabs you.

We often hear the phrase "live and learn." I'm not sure it shouldn't be the other way around.

The Problem Statement: How are the 10 digits placed on 2 separate cubes in order to be able to denote every day of the month? Have a solution? E-mail it. (See* ***Appendix 2* *for contact information.)*

For a discussion of various solutions to this problem (and its answer), see ***Calendar Cubes and Gumballs...*** *in Section 7B.*

SECTION 3

Perspective – Use It or Lose It

INTRODUCTION

Not long ago, my wife and I re-watched the old movie *Fiddler on the Roof.* For days afterwards, I caught myself singing the "Tradition" song. In the movie, the father's traditions are sorely tested.

This is exactly how it should be with perspectives. We should have occasions to have ours tested, shaken, and/or shifted, or at least examined, to see if our beliefs still match our practices and/or our changing times. Sometimes, a different view of things can help make a breakthrough.

Often our perceived ideas of an ideal education (as varied as those are) don't translate well into the classroom. Somehow, over years of "doing it the same way," we lose our perspective, and by the time we realize that our perspectives have been tested, the damage is often done.

This section is about *perspective,* but it also feels like another "spitball from the back row." This wasn't my conscious intention, but some of these essays *could* be construed, in some manner or other, as going against what might be considered the current, or even new, conventional wisdom—or at least standard practice in a certain area of education. A difference in perspective here can at least open the doors for discussion, whether we all agree or not.

THE CHAIN OF INFLUENCE

We're all familiar with chains of command. Many of us deal with them all the time. We know to whom we should take a question or problem, and we know to whom it isn't kosher to approach with the very same issue.

This chain of command exists in school systems as well, of course, but, for my money, it manifests itself in an interesting way, and creates some interesting consequences. More in a minute.

Let's start with the classroom teacher. This is over-simplified, of course, but the teacher's immediate "boss" is usually the principal, who in turn answers to the superintendent (with perhaps another level or two here at bigger schools), who is responsible to the local school board. And all of these entities, at least in theory, answer to the community, who has elected the Board. This, roughly, is the chain of command.

With various exceptions and subtleties along the line of course, as one moves "up" the chain of command, each level gains more and more power over what happens in a school system in general, especially in terms of policies, procedures, etc. This is normal, and certainly not without its strengths. A lot of good things get done efficiently that way.

But there's an interesting feature about a school's chain of command. As one moves back "down" this chain, one gradually returns ever closer to the actual classroom itself, where the rubber meets the road, so to speak. This is where the real *learning* can happen, and where real differences can be made on a day to day, even minute to minute, basis. If moving up this chain is called the chain of command, then moving down it should be called the chain of influence. One gains power moving up, but one gains influence with students moving down. And, moving down, it ends "at the top" with a full circle return to the teacher.

This creates, however, some interesting consequences. The further away one gets from the teacher, the more power that is gained. But this leaves the teacher in an awkward and often absurd position: The one person with the most influence and knowledge about what happens in the classroom and how/when to effect genuine learning, is, tragically, often the person who has the least real *power* over important system-wide decisions.

There's a lot of talk about "bad teachers" these days. We seem obsessed with the idea of getting rid of them, and we do so, in my opinion, at the risk of losing *much* perspective. I say this carefully, but I'm convinced that much, if not most, of this talk is pure (often politically-influenced) rhetoric and misplaced finger-pointing. For every "bad" teacher (how that is determined or identified is a topic for another time), there are literally handfuls of good teachers, making a difference, and doing so against very tall odds and less-than-ideal conditions.

Please understand, I am *not* against "accountability," and I am *not* defending "bad teachers." But I *am* saying that what we tend to do (whether intentionally or not) is to give teachers more and more responsibility and less and less power to go with it. We overly blame them for the "bad things" that can happen while we simultaneously hamstring their ability to fix the bad things and create good things.

Wouldn't it seem to be helpful, aross the board, if we paid more attention to the chain of influence in schools, and made it more important in the chain of command's decision-making structure?

EDUCATION IN THE INFORMATION AGE

I thought I was being purposely outrageous. I didn't realize I was being a prophet.

Early in my career of working with future teachers (decades ago), I can remember saying to them, "What will you do when there is a wrist-watch device that will instantly give you all the *information* you want? How will that affect what and how you'll be teaching? Will you, the teacher, become obsolete?"

The idea, of course, was to get them to step back and start thinking about the fact that "becoming educated" is much more than—indeed entirely different than—"gathering information." At the time (late 70s, early 80s), such a question was a "thought-experiment"—and an effective one at that, most of the time. The "shock value" of an instant-information environment, especially in those times, provided an opportunity for future teachers to at least be thinking about some important types of things that could be happening in the classroom, other that rote learning, or memorizing, or gathering information.

Now, as we are all aware, such a future is the present. Such devices now exist, and even if your "device" isn't exactly a watch, it is just as accessible for most of us. For anyone who has a computer—or now a smart phone—gathering information is a relatively easy task. And as you think about it, this has drastically changed our lives and our habits.

By the way, interestingly, this almost makes it harder for future teachers to ponder these ideas NOW than back in the "wristwatch thought-experiment" days. Now, students come *straight from* this environment, so they haven't even thought about the implications for their future classrooms. And they have no alternative context to highlight the problem.

The cartoon strip *Doonesbury* had a marvelous strip a few years back (6/26/11) that captures this dilemma.* The next-to-last frame there sets up the humorous punch line in the final panel by asking essentially "what does it mean to be a student?" Ah, but the more crucial, system-shaking question is really, "what does it mean to be a teacher?"

Understandably, but sadly, our educational system is well behind the curve in developing an answer to—indeed even re-asking—this question. Much (not all of course) of our curricula *at almost any level,* is still highly information-based. That's not necessarily all bad,—and again, it's certainly understandable—*but,* it seems that none of us is pondering "bigger picture" questions such as the following:

Now that we are well into the 21st century in a highly technological society, what (and how) do we want our kids to learn?** Are we teaching kids things they could as easily find on their own when they need them? If so, how shall we use valuable class time instead? What/how *should* we be teaching to prepare students for their adult lives in this rapidly changing technological world?

And more drastically: What would be an equivalent outrageous "wrist watch" question that will come true in future and are we preparing them for that scenario, as well?

These are highly "loaded" questions, of course. And one person's answer may be another person's heresy or nightmare. But, if we aren't even *asking* these questions, how can we go about answering them? Are we willing to let our students continue to "learn" accessible information?

Henry Taitt once said "Tell [students] what to think, and we make [them] slaves to our knowledge. Teach [them] how to think, and we make all knowledge their slave." The quote is a little narrower than the thrust of the column, but it makes the point: *Which* do we want to do, and *which* are we actually *doing*?

**The cartoon may be seen by going to https://www.gocomics.com/doonesbury/2011/06/26*

This question is also explored in* ***What Do We Want Them to Know, *in Section 1, as well as other less obvious selections.*

For later perspective on this same topic, see ***Where Do We Go From Here?*** *in Section 8.*

TWELVE MONTH SCHOOL YEAR

Well over a year ago, I responded to a "Letter to the Editor" that appeared in the paper. The letter was bemoaning that students had missed six days during the school year, due to snow, and then went on to argue for year-round school. The letter was entitled "Students Need More Education, Not Less."

I'd like to take some of those thoughts in my response then, and recycle some of them into a column, since the ideas there will piggyback onto other similar explorations that appear elsewhere in this book.

We should tread carefully and sensibly here, of course. Who would ever argue that students need *less* education? But let's not miss the forest for the trees. The bugaboo is making sure we focus clearly on the product we're delivering in these rapidly changing times. Only then can we move to discussing whether we want to deliver more (or less) of it in a predetermined time span.

The author himself made this exact same point in his letter, perhaps without realizing it. "Public education seems to be struggling to survive. Throwing more money at the problem won't fix it until there is agreement on what's needed."

Amen! I couldn't agree with this more. In that case, let's have a good, rational, public discussion, with all constituencies represented, about "what's needed." More to the point, let's join the discussion that's already occurring, and, in the process, *let's listen to the teachers*, who are on the front lines and who deal with promoting learning every day.

Let's not jump straight to the assumption that more time in school (or any other "easy solution") will fix the problem. Going to school more won't necessarily make one more educated... any more than going to church more will necessarily make one more spiritual.

I'll say this carefully, but I've *always* thought the longer-school-year argument *by itself* was a lot like saying "My car is running rough, so I need to drive it more miles." Shouldn't it be clear that "what's needed" is not *necessarily* more quantity but more *quality* in what we do, whether it's church, school, career, or whatever?

Yes, let's all pitch in and help educators as they try to make *authentic* learning easier to attain in the classroom. Let's all help find ways to connect "the curriculum" to "the outside world" in ways that make sense to students, so they want to learn, and understand *why* they are learning. And let's *support* the classroom teachers, whose daily mission is to help achieve those very goals.

These goals, especially today, are *not* easy ones, and they're made more difficult by not realizing how legitimately complex they are, and how difficult they are to achieve and sustain. So, let's join and contribute to that discussion with open minds. It's important. Let's *not* get sidetracked with whether we should do the same old thing more or less often. It tends to miss the point.

DOCTORS AND EDUCATORS

It was a beautiful summer's evening, with a view of the lake, and the sun preparing to take a dip into it. We had joined friends for dinner, and other friends joined us for dessert. The conversation was lively, and eventually turned to the national economic troubles and then to health care. At one point, one of our friends said, "I hear now that they're going to pay doctors by whether or not their patients get better. More if they do, less if they don't."

My first thought was to agree this seems stupid and my second thought was to wonder if this was another urban legend rumor. Those were my *thoughts.* Instead, what I *said* was... "And yet they think this works for teachers."

Not surprisingly, I got some interesting reactions (like you perhaps just had?) Then, because they're friends, there were a couple of polite "Hmms..." and the conversation moved on. As I had no real wish to pursue the point (nor to be later killed by my wife), I let the discussion meander away.

Not much later, as the group was deciding that no one in Washington knows what they're doing (have you guessed that wine was present and flowing?), someone said, "After all, there are only two *doctors* in the whole congress." This time, I didn't say anything (I think I was too busy nursing the pain in my shin where my wife kicked it), but can you guess what I thought? I'll bet that's two more than the number of educators!

I've been thinking about these parallels ever since. What do you think? Perhaps the doctor/educator analogy is a little far-fetched for you. It might even be for me, but, then again, I'm not so sure. It makes me wonder. Just because a doctor "treats" a patient does not always mean the patient will be cured, for a lot of reasons. We all know this (whether we think Congress does or not). And while it is true that some "good" doctors can cure patients that others might not (same with teachers), it is also true that even the best doctors have patients that die. Just because a doctor "treats" doesn't necessarily mean the patient will be cured.

So why do we think that if a teacher "teaches," the student *should* automatically "learn?" Is the difference in our views of "medicine" and "education," or is it something else? Perhaps we see a fundamental difference in the arts of teaching and healing. (Maybe we think healing is an art, while teaching is a craft). Perhaps we think that while it takes a lot of training to be a doctor, being a teacher is easy. Perhaps we've never thought about it.

The point remains: teaching and learning don't *always* go together, any more than treating and healing. Maybe we think they *should.* But they don't. It's harder in these tumultuous times for them to go together than it has ever been. This is true even for the best of teachers, and it is true at every level. So, I tend to frown (and you should too) when someone (usually a legislator) begins talking about "fixing" our schools, and "making teachers accountable." Please understand: I am <u>not</u> against accountability, (however we define it) and I am <u>not</u> defending "bad teachers" (whatever *that* means). It's just that it's not that easy. There are always several very complex issues involved—that's the nature of educational issues.

We'd be well-served to be warier of folks with quick-fix solutions or simplistic political bandages. Anyone who thinks they know how to "cure" our educational weaknesses with a piece of legislation probably thinks that doctors of terminally ill patients should not be paid.

HOW EFFECTIVE ARE SOME RULES?

I'm still an optimist. I like to think that every educational policy which sprouts a glaring weakness is one that at least *started* with good intentions. At the same time, we all know of the paving of that infamous road south, don't we?

The e-mail from a student that day was only two lines long—no subject line and not even signed. It said, "Hey I was wondering when our next test is? And can you tell me how many days I have left to miss class? Thanks." It's only been after years of receiving similar notes that I've learned that it IS possible to laugh and cry at the same time.

You might have guessed that I had announced/reminded (aloud and with a PowerPoint slide) when the next test was in *every class period* for at least the previous two weeks. And this student, whom I'll call Sally (name and possibly gender have been changed) was almost always in class. As I think about it, the question about the test doesn't really bother me, but it connects to her other question—and this column's topic—really well.

I'm more intrigued by Sally's view of attendance, and I'm wondering if that attitude, as stereotypical as it may be was (certainly unintentionally) fostered by this institution, which only allows a small number of "misses" for a class. On the surface, increased attendance is a noble goal, as we'd all like to believe there should be an obvious relationship between attendance and learning. As a *policy* however (which *is* different), required attendance leaves something to be desired.

Is there, in fact, a connection between attendance and learning? Usually, yes, one would hope. But not always. In Sally's case, for example, it's blatantly clear that Sally missed that connection. Notice: Sally was in attendance for several of the classes in which the test date was mentioned, and had clearly missed the information. Her usual attentiveness, or lack thereof, in class suggested to me she that had missed much other material, too. (This is not necessarily a criticism. She "knew" much of the material. Indeed, that was part of the problem.) And further, she wasn't trying to hide her absence planning.

From my perspective, Sally was often not there even when she was attending! If she didn't want or need to pay attention (and was not distracting others), I wasn't going to waste class time trying to make her. She's an adult. And I wonder, if she wasn't paying attention, why should she be required to be there? There's a compelling case that she and I (and the class) would have been just as well served if she weren't.

There are *many* other examples of similar "requirements" or rules that don't really achieve their purpose, and I hope to discuss more, including one or two of my *real* pet peeves. But, as always, let's keep the perspective clear. The subject here is not necessarily to debate mandatory attendance policies. Instead, I'm trying, as always, to underscore the fact that educational issues aren't easy, and policies—especially those that try to govern behavior—aren't always as effective as they might seem at first. Sometimes less is more.

BEWITCHED, BOTHERED, AND BEWILDERED: A RIDDLE

—

This story really *did* happen, but I've never been able to fully get my head around it. Was I the "victim" of a good-natured student joke? Because the obvious explanation is just too hard to believe. Perhaps you can help me decide.

Not long after my retirement, I began to miss interaction with students, and agreed to teach an introductory algebra class at a local institution. It met twice a week from 5:30-6:45.

One of the students was a pleasant young man just out of high school—we'll call him Jeremy—who was personable, caused no trouble, but never seemed too terribly interested in the class or the subject matter. He had all the obvious symptoms of taking a class to satisfy a requirement and move on.

Jeremy always attended with his female friend—we'll call her Jane—who exhibited similar symptoms, but not *quite* as pronounced. They seemed to try help each other stay motivated in a subject they'd seen *some* of before. They were a *team*, for better or worse. All of which is fine, of course.

Jeremy and Jane almost always arrived with about 3 or 4 minutes to spare and were usually ready to go when class began. They always tried to (appear to?) exhibit effort during required-by-the-institution self/group work periods, but the minute I allowed the class the freedom to leave if they had no more questions, they were usually gone. (Others stayed, worked, and asked questions.) They seemed to have this sense of "duty", of wanting to do the proper outward thing, but I wouldn't call it dedication.

One afternoon, I wandered into the classroom about an hour early, as usual, to get set up and be available for questions, and there sat Jeremy and Jane. They were in their seats, but had no work out and did not appear to have questions. Early arrival was not their pattern, and I was surprised, but I didn't think too much about it. I greeted them and went to set up.

Later, I wandered over to chat with them and see if they had questions. It was then that the reason for their early arrival was explained to me. One of them—I forget which—told me, "We have to leave class early this evening, so we came in early to make up for it."

I'll let that sink in for a minute.

For a lot of reasons—decisions I made instantly—I decided it best to accept that at face value. I now wish I'd have explored more. Sure enough, they later left class considerably early.

This incident started as a "funny student-story" for my storytelling archives. As time passes, the humor has changed to puzzled intrigue. In the first place, there was this fascinating sense of "duty." They honestly seemed to feel some kind of need to conscientiously replace the minutes they were going to miss in class by being there in the classroom early. (Was that only to appear conscientious?)

But of course, that pales alongside the obvious question. Did they *really* believe those minutes sitting in the empty classroom replaced the experience they might have gotten during the class itself? I find that *so* hard to fathom that I continue to wonder if they were "pulling my leg," all the while with straight faces. I almost prefer the latter, because to believe they were serious leaves so many (*too* many) loaded questions unanswered—still.

Should I laugh at this story, or be generally troubled? So far, it's been both. Any thoughts?

YOU SAID, "HAVE FUN!"

I used to teach a capstone course for future middle school math teachers. At this future-teaching level, it becomes important to have a broad taste of the development of mathematics through history. But students don't want to sit through a stereotypical two-week "unit" on math history. And, since I'm not that anxious to do that either, I developed an alternate plan.

I'd ask the students to write a paper on an interesting math topic or person of their choice. With good guidance, some logistics maneuvers, and some creative grading twists, this collection of shared papers usually achieved the desired purpose of giving them a broad background of early mathematics development.

At the start, students *always* worried about "what I wanted" for the papers. How many pages? What about references? Would writing/ spelling count? All the typical "student" questions.

It was (understandably) hard to allay those fears and get them focused on the bigger picture. I always suggested they find a topic they could get interested in (I gave lots of suggestions, individually and collectively), and find a way to share it with their fellow students in a way *they* would find interesting and would want to read themselves. Creativity was allowed, even encouraged. My favorite phrase was to try to "have fun with the paper." I regularly had some *great* submissions. I was almost always pleased in general.

One semester, a student turned in a paper that was considerably short of the mark. It was sloppy and felt thrown together (which is usually a good sign that it was). It was clear that she hadn't gotten much from the paper, nor would her classmates. Sub-par submissions weren't altogether unusual, as one might expect. What *was* unusual was what happened next.

My written comments to the student, and the "grade" she received reflected her lack of effort and/or "results." She was unhappy about this and wanted to talk about it. Ironically, she didn't dispute her lackadaisical effort or even her (lack of) learning. Instead—I swear this is true—she said, "You just said to have fun with it. You didn't say it had to be good!"

It was one of the very few times in my career that I was literally speechless. I didn't know whether to laugh or cry. I honestly don't remember what I said to the student. I suspect that, whatever it was, it didn't help, and she didn't get it.

But that's the point, she didn't get it, in general. It later dawned on me that her view of getting an education was about checking things off a to-do list as rapidly as possible. It had nothing to do with *expanding her knowledge/preparation* for teaching or *growing* in the subject. Indeed, it had little to do with *learning* at all.

I don't necessarily blame this student. I confess this view of education is prevalent in most college (and other) students. Ironically—and tragically—I saw this attitude the most in future teachers, which this young lady was. What does this say about the kind of environment that will exist in those teachers' classrooms?

Indeed, what does it say about the attitude that must have existed in at least *some* of these students' classrooms as they came through the system at all levels?

I mentioned that I don't necessarily blame the student. Moreover, I honestly do not blame the teachers either. If there is "blame" it should go to the nebulous system at large, and to us. *The system* has subtlety created a dynamic where "getting educated" means "finishing courses" and "getting the piece of paper," and then *we* assume, or agree to accept, that this corresponds to meaningful learning.

Another of the big tasks facing education today is not just pondering how to re-inject the goal of meaningful learning into the process. It's to awaken and become aware that this goal has already been lost in the shuffle.We can't do the former until we recognize the latter.

THE MISS PERCEPTION PAGEANT

—

Years ago, I gave the keynote address at a statewide conference for math and science teachers. I called it "The Miss Perception Pageant" and spoke of some of the misperceptions that exist in the general public (and with our students) about these two important fields, both in general and as they relate to education and our teaching of these topics.

An interesting by-product of preparing for that talk was to discover yet *another* very general—and contradictory—misperception I hadn't thought about. This discovery is best told as a fun story.

Because I was going to be speaking partly about misperceptions related to the public, I decided to visit with my most-easily-accessible "(wo)man on the street"—my wife. Recall that my wife, before she retired, was a highly successful, articulate, and inspiring business woman, but, like many of us, she is not the most comfortable with topics in (or about) mathematics.

After some coaxing, she agreed to answer some broad questions. Here's a pretty accurate transcript of the conversation:

Larry: So, how would you define math?

Pat: (after some thought): Understanding numbers and their relationships.

Larry: Okay. Not bad. So, where would you put algebra then?

Pat: In the trashcan.

Larry: (pausing to grin and digest...).

Pat (not waiting): No, really. It's meaningless. In the trash can."

Larry (still laughing, but trying to regain control): Hmmm. Well, okay, how about geometry then?

Pat: Oh, well, now. I liked geometry... Okay, maybe you better add angles and shapes."

By that time, we were enjoying ourselves, but I wasn't sure this was going to help me. I switched to the field of science while I still could.

Larry: Well, then, how would you define science?

Pat: Understanding how the universe works.

Later, as I thought about this conversation, a couple of things dawned on me. I believe that Pat had been the *perfect* spokesperson to represent the majority of "mainstream" perceptions of math and science.

On the one hand, many of us tend to *underestimate* the value/power of mathematics when we view it as "numbers and relationships" *only*. When we think of math that way (as in "do the math"), we cheat ourselves. It's what I mean and have meant when I continually say, "Math is not arithmetic—only."

At the same time, I think many of us run the risk of having almost the opposite misperception of science. Except for that fascinating subset of folks who are "science doubters" and fact-ignorers on seemingly all belief-threatening topics of religion or politics, I don't think we under*value* science. But I think we sometimes overestimate its *scope*. While the work of science over the decades has been amazing, it does not "understand how the universe works" on a comprehensive scale at all. Obviously, an understanding of how things work is the goal of all branches of science, but that is always a work in progress, and there will always be scores of unanswered questions in every field. The "understanding how the universe works" is a process, not an acquired destination. A scientist is an explorer, not a possessor of all knowledge.

This common disparity of views certainly has implications for education and teachers in both fields. If we did a better job of shining a light on what math *can* do, would we decrease some of the anxiety? If we did a better job of putting perspective on how science and scientists work, would we decrease some of the fears?

SECTION 4
Life and the Real World

INTRODUCTION

As we prepare students for "the future," it seems to me that our perspectives can often get out of balance in *both* directions. Sometimes we tend to miss the forest for the trees: we get so focused on the "academic" nature of what happens in schools, that we lose sight of the fact that we're preparing students to use these academic (and other) tools in the classroom of life, out there in "the real world."

Other times, we can tend to get so hung up complaining that we're not preparing students for the "real world," we can forget that this is a huge task with a nebulously articulated set of sub-goals along the path. And that path has no easy road map. We can't just hand them a "prepared to tackle life" certificate.

There's a delicate balance, here, with several issues to consider. Some of those issues are subtle enough that they can get lost in the shuffle, if we're not careful. So they're worth noting, and discussing. The columns in this section are selected to highlight a few of those issues.

TEACHING LIFE? A TOP TEN LIST

For a week or so not too long ago, one of the top educational news stories revolved around the Department of Education's two gaffes of spelling and grammar when they posted a quote by W.E.B. Du Bois. I think the temporary uproar may have eclipsed a focus on Du Bois' quote itself, which deserves noticing: "Education must not simply teach work—it must teach life."

We've all heard variations of this theme, from "Teach a man to fish..." to my own favorite that says, "We must prepare our students to make a life as well as to make a living."

As always, there'd probably be general agreement with the spirit of this *thought*, but perhaps disagreement and controversy deciding *how* we do this. So, let's re-focus the question. Instead of "how to teach" life, let's ask "what should students learn (somehow, sometime, some way) about life?" That's just as subjective, of course, but perhaps makes for more tangible discussions. It's worth thinking about.

Understanding *both* the subjectivity and the risk, I'll tentatively offer my own Top Ten List of Life's Lessons it'd be wonderful if our students knew as/before they left school:

10. The arts are as important as the sciences in preparing us for "living a life." So is the study of history, so as not to repeat mistakes of the past.
9. Reasoning and rational thinking skills are crucial in all the roles of a good citizen. (These are the broader goals of most curriculum topics.)
8. Honesty is still the best policy.
7. "Every action done in company ought to be with some sign or respect to those that are present." George Washington
6. Disagreement is with *ideas*, not *people*.
5. Compassion and compromise are signs of strength, *not* weakness.

4. Bullying, in any form, is a sign of weakness, not strength.
3. If you wouldn't say it directly to a person, why would you say it online?
2. A person's race, gender, spiritual path, or sexual orientation has *nothing* to do with that person's innate ability to succeed at what they do, and/or to contribute to the good of society. (Even if you/we view the religion as heretical and the orientation as sinful.)
1. By and large, "ordinary people"—our neighbors, in every sense—are decent, good humans. We (all of us) are worth the effort needed to *listen to* each other, talk (even debate) with each other, and mutually strive to reach solutions for the good of our country, our communities, and our lives.

Ah, yes. I'm sure I'll be tried and convicted of being hopelessly idealistic. Nonetheless, many of us look around us these days and wonder how the heck we ended up here, in these uncertain and angry times. Could it be that we're not really doing a respectable job of teaching, and then following, Du Bois' advice?

One final, important note, which I'll say carefully. I'm not sure these things should be *taught* in school, as much as they should be widely *reinforced* there. Hearing these maxims in school should elicit "well, duh!" responses from students, who have already heard them, in various phrasings, at home and in society, and more importantly, seen them in action.

Obviously, we should continue to prepare our students to succeed in whatever careers they choose. But we should also prepare them for living a life, not only in their careers, but in society as well, and we need to think about what that means.

HOW DID YOU PREPARE FOR YOUR CAREER?

"Did you know at HS age what you wanted to be?" This question was in the middle of a longer thought-provoking email discussion with a reader. I was intrigued for a variety of reasons. The question seemed to lead down several trails. Yet, I think these paths all merge at the end.

1. The first trail is a memory. I sometimes look back on my own "career preparation" and marvel. I agonized very little over that path, for almost embarrassing path-of-least-resistance reasons. Painful as it is to admit, I proceeded through college in a state of blissful "un-concerned-ness." That I should have somehow ended up doing what I loved, in a subject I loved, at a level that suited me—and getting paid for it—is often amazing to me. I look back and wonder who wrote that script. I almost feel guilty.
2. On the far other end of the spectrum, of course, there are students who want a challenging career, but have no idea what they want to do or be, even well into college, and they *are* concerned, even panicked. I used to see it all the time as an advisor and professor. Students feel so much outside (or even in-school) pressure to pick a major, or a career path. I believe we need to accept that uncertainty and help them through these times. Indeed, this should be okay, even normal! It should be a time of exploration and dabbling, *not* a time of being forced into choices too early, either by course requirements or "declare your major by such-and-such" rules of some kind, even if it delays their (or someone's) plan for graduation-in-four-years. (I know, I know. They should then be reminded this could be costly financially. I'm not ignoring that. But there are cases where it would be well worth it, even financially, in the longer run.)

3. And, there are also those students who know exactly where they want to be at sixteen, and that's *out* of school. The traditional wisdom is that dropping out is generally an unwise decision. And I don't disagree in general. But I've always thought—and I say this carefully—that "dropping out" should be made slightly easier (while being counseled) if it was also somehow made easier for the student to "drop back in" from weeks to years later, when they discovered they wanted to. They would come back more motivated, with better perspective. Perhaps the new Adult High School movement (in my area, and elsewhere) can help in this regard.

These different situations lead to one over-riding and fascinating question for me. How do we most efficiently prepare the biggest percentage of our students for the futures ahead of them, when they don't yet know what they want to do? How do we prepare them both for making a both a living and a life?

How do we give them choices, and still provide depth in the right places? How do we walk the fine line between "broadly preparing them" in general and "over- or under-preparing them" in specifics? How do we give them, as one example, enough math & science to help them be well-prepared citizens, without stuffing them with more esoteric "requirements" if they don't later pursue a career in math and/or science?

These are age-old and understandably complex questions. They remain part of our constant "what *is* education?" theme. But today's society and circumstances require ongoing, fresh, creative thinking about their answers.

BACK TO THE FUTURE

Back in Section 2 ***(Searching for Right Answers)***, I related an incident that occurred "back in the day." I had made a careless arithmetic error that affected my test score and, while I had worked the problems with the "correct" procedures, my teacher had told me that "in real life, you don't get partial credit for a building that falls down."

Let's leave that incident itself (and the participants) back in the '60s, but move that thought into the present. With an eye on the future, let's explore the broader issues in play, along with some others. For those broader issues and dilemmas are still very much alive.

As an educator it seems that the basic underlying question is this: How do we prepare our students to learn the skills they need for their future professions, and still allow them to make the necessary mistakes that allow them to *learn* those skills in the process? Where do we draw the line?

Clearly, we want to prepare our students for "real life" situations, and the "perfection" those situations require. (If one builds a bridge, it must stand up to decades of brutal treatment.) Yet our students are NOT in "real life" yet, and they are often far from having developed the skills they will need when they arrive. Is that not what "school" is for, after all?

These important questions are not restricted to any one discipline, of course. How do we teach youngsters to ride bicycles without knowing they will spend some time losing their balance and skinning their knees? How do we hone the skills of a young future Hemingway without putting up with—and correcting—some boring and pedantic writing in the process? How do we encourage a future Yo Yo Ma who has just picked up his/her first cello, and who will make some interesting sounds in the beginning? The analogies are abundant.

As we prepare and work with our students in all their disciplines, it makes sense that we must be patient and allow, even encourage, the missteps that will occur along the way. It's all part of a learning process that prepares students for "the real world." Part of what makes a good teacher is knowing when and how rigidly to walk that fine line for the benefit of all.

Interestingly, though, there is another side of the same coin. There are often places where we are shortsighted in the other direction, and *could* do a better job of preparing for "the real world" earlier than we do. In my own discipline, for example, I have often wondered how we prepare students for the "real world" where folks work in teams, use technology (both for communication and computing), and often extend projects over weeks at a time, when we force them to do arithmetic problems by themselves, using paper/pencils only (no technology), and operating within a time limit. It doesn't make sense. Other such examples abound.

As always here, the ongoing task of educating our youth for the future is a long-term *process* that is a delicate tightrope walk. We must simultaneously keep our eyes on the future product, while focusing on the present, and working with students as they are when we get them. It's a delicate and dangerous balance. It's one small part of what makes education both a challenging and a rewarding endeavor.

FIGHTING CANCER AND THINKING CRITICALLY

On a fine almost-Spring evening not long ago, my wife and I attended a marvelous event in Kansas City which contained some very powerful educational food for thought.

William Jewell College hosted its Annual Achievement Day Dinner, and the featured speaker was Dr. Siddhartha Mukherjee, physician, assistant professor of medicine at Columbia University, Pulitzer Prize winning author and, as it turned out, an excellent speaker. Dr. Mukherjee won the Pulitzer Prize for his first book, *The Emperor of All Maladies: A Biography of Cancer.* This book was also the basis for a Ken Burns PBS documentary, and helped place Dr. Mukherjee on *Time* magazine's list of 100 most influential people. His newest book *The Gene: An Intimate History* became a number one *New York Times* bestseller.

If you're wondering how the topic of cancer relates to an educational column, read on.

In an oversimplified nutshell, Dr. Mukherjee spoke of how *interdisciplinary* the battle against cancer has become. He spoke of how researchers need the help of so many others: scholars from the information and technology sciences to help organize and do mappings for the overwhelming amounts of data collected; social scientists, as they examine the way patients interact with medicines and their own belief structures; linguists, since communication between cancer (and even other) cells often resembles a language; nurses, as they are often the crucial caregivers and inspiration for patients, and patients themselves, as they work with their doctors to exchange information. He also included molecular biologists, chemists, and others. It was (and is) a fascinating overview and perspective.

As he finished, Dr. Mukherjee noted that cancer is just one example of how the increasingly complex problems of today require more and more interdisciplinary approaches from so many fields. He spoke of how these various fields need to be populated with professionals who have been trained as critical thinkers and who can also see beyond their own areas of expertise. And, of course, how they need to be able to communicate meaningfully with each other to jointly attack problems.

Just as impressive, earlier in the evening, were the brief remarks from William Jewell's new President, Dr. Elizabeth MacLeod Walls. These remarks served as both an introduction of the speaker, and, as it turned out, a perfect segue to his remarks as well. She spoke of the fact that Jewell has long been known for developing critical thinkers of its graduates. She also highlighted Jewell's new Brand Mission: *We are critical thinkers in community pursuing meaningful lives*, a statement that says so much in so few words. If you've read many of these columns, you'll recognize some favorite themes dancing in and around that sentence.

The evening and the two sets of remarks combined to serve as an enlightening reminder of some of our higher, less esoteric, goals and values of our educational process, from kindergarten through college/university. And these goals and values are not restricted to higher education parts of that process.

Critical thinking, interdisciplinary thinking, cross-pollination and fertilization, sharing of ideas, effective communication, tackling (and eventually solving) major problems, leading meaningful lives—such an inspiring list! These things contribute to our growth as humans and citizens. And they contribute to our ability to help solve the problems of our own lives and of society. They are a grander part of becoming educated.

FUNERALS, FIDDLING, AND THE FUTURE

I had attended the funeral of a former colleague who was a longtime music educator at a former institution. I was impressed and inspired by the folks who spoke and shared stories about my former colleague. Most of these speakers were former students now doing impressive things in all realms of the music world.

When a funeral is done right, it seems to me, one leaves the service inspired by the deceased's life, but also reminded of some important perspectives about life in general, and in this case, education. I know that as I left, I was indelibly reminded of three things.

I was reminded, first, of the value and impact of good teaching and a good teacher. Students who spoke of their mentor's influence in touching—and funny—terms were clearly now passing along those same influences in multiple ways to those for whom they have become mentors. And, of course, this impact will continue for untold generations.

Second, I was reminded of the importance of music in particular, and the arts in general, in our lives. Which of us does not have our own music (of *whatever* style) that speaks to us, that moves us, that energizes us, that supports us?

Finally, of course, I was then similarly reminded of the resulting importance of music in particular, and the arts in general, in our educational systems. Think of that music that moves us, whether classical, rock & roll, or other. Where do we think those musicians first discovered and then began to cultivate their craft? Do we think they suddenly decided, after graduating from high school, "I want to play the saxophone?"

We must somehow, as a society, find ways to slow the elimination of music in particular, and the arts in general, from our educational curriculum. Or we must find a (financial?) way to allow schools to continue to provide these crucial elements of education for students to discover. Or, we must find ways to do it by ourselves.

Along those lines, meet Ozark Mountain Music, Inc. Their mission is to pass the traditional music of the Ozarks on to other generations of fiddlers. Ozark Mountain Music runs fiddle camps each summer, and after-school programs during the school year. The Possum Holler Fiddlers are the showcase fiddlers of the program, performing at various kinds of events throughout the year.

The program receives some support from the Missouri Arts Council, but it is the brainchild of Bob and Karlene McGill. They are doing wonderful things with the programs *and* with the lives of the young fiddlers, some with exceptional talent, previously unknown. Discover much more at Ozarkmountainmusic.com, including pictures and contacts.

Traditional Ozark fiddling is not likely to be in the curriculum of many schools, so here is a group doing something about that on its own. Will we eventually need the same type of groups in various locales, for the more traditional music and arts programs?

All of which leads us to the future. How do we manage, in these challenging times for education, to preserve environments which sustain excellent teaching and mentoring for our students? How do we manage to preserve music and the arts in our schools? We must find creative ways to maintain these crucial features of our educational systems. We must do this for our society, for our students and for *their* futures.

ATHLETICS AND THE THREE A'S

In the previous essay, I suggested briefly that we all suffer when "the arts" are neglected in our school curricula. (An even stronger case is made in ***Pluto, Scientists & the Arts***, Sec 10).

For this piece, I'd like to live a little more dangerously, and take a first quick look at the role of athletics in our lives and our schools. For the purposes of this column, I'll focus more on the influences in my own personal life. We'll expand that and make some further connections in ***The Three A's—Part 2***, also in Section 10.

Like it or not, for good or not, athletics does influence our lives, especially as participants, fans, parents, teachers. For my own part, several personal scenes immediately pop to mind. To me, these are all related.

SCENE 1: Greencastle, Indiana, late 60s. I'm just starting my nomadic sojourn through graduate school(s). My wife is teaching sixth grade and is expected to coach her class's football and basketball teams. I gladly offer to help.

We have fun with the kids and I have fun playing coach. Unfortunately, this includes questioning nearly every officiating call made in every game we played. At times, these sixth graders are more mature than I am. My new wife decides to write me out of the will.

SCENE 2: Southwest Missouri, late 70s. For an autumn or two, I decide to officiate some local high school football games on Friday nights. Wow. Those coaches and fans (and parents!) can sure take these games way too seriously. Was I this bad in my Greencastle days?

SCENE 3: Southwest Missouri, Winter '93 or '94. As a good parent, I attend "our" high school's basketball game to cheer on the team. I get caught up in the excitement, as do others, and we loudly begin to help the two officials, who are clearly missing some obvious calls. Soon, those crazy fans from the other team, who have the same poor vision as the refs have become part of the enemy.

In the middle of one of my yells, I wake up, catch a glimpse of us, and wonder if some of those fans might consider us as obnoxious as they seem to be. (And if the officials might view all of us the way I viewed fans in Scene 2.)

After the game, I notice the players and coaches all shaking hands, and I'm thinking that might not be such a bad idea for opposing fans as well. The players have left on good terms—have we?

SCENE 4: Victoria, British Columbia, Fall 1994. At a math teachers' conference, a *music* teacher comes up to visit after my talk. He says he attends conventions of other disciplines to get other clever ideas. WOW. Anyway, among other things, he suggests that "the three R's" ought to be "the three A's:" Academics, Arts, and Athletics.

The Three A's. I like that. We'll follow up more in Section 10, as mentioned. In the meantime, beware. This is *not* meant to be a "pro-athletics" column, *per se*. As a participant, coach, official, parent, fan, and stuffy academic professor, I've witnessed the addicting lure of the dark side of athletics.

But it is to say that athletics have played a role in my life, not just in school-related ways, but as a vehicle for growth, development, and perspective as well. The coaches like to call it "character building," and I don't think that's far from wrong.

A lingering question: As we continue to prepare our students for life, should we acknowledge, encourage, and stimulate the role of athletics (and the arts) in their lives—or at least prepare them to examine that role for themselves? And by so doing, might we not make them better participants, patrons, fans, and even adults?

SECTION 5

The Magic (and the Magicians) of the Classroom

INTRODUCTION

When it comes to schools, at least, most learning starts in that mysterious place called "the classroom." Here is where things happen, for better or for worse. Here is where the rubber meets the road. For the most part, teachers have control of what happens here (see the next section for issues where they *don't* usually have control) and they are often masters of making decisions that greatly influence their students short- and long-range learning.

This section will be not unlike Section 3 (*Perspective*) in that, as heroic as teachers are, and as often as they are perpetrators of miracles, we often fall into habits that perpetuate themselves past the point of usefulness. (Or at least I did—maybe I should only speak for myself?)

In that vein, then, this section relates directly to three important entities: A) Teachers themselves, B) the environments which they preside over, and C) the perspectives that drive some of their classroom decisions.

HERE'S TO THE TEACHERS!

Everyone who remembers his own education remembers teachers, not methods and techniques. The teacher is the heart of the educational system.

Sidney Hook

This is true, isn't it? Aside from our memories of school events, friends, social fiascos, and such, our thoughts back to our younger classroom days are dominated by memories of teachers and not so much the pedagogy they used, or the subject they taught, right?

Discussions about teaching these days seem too often centered on issues like "accountability" and "getting rid of bad teachers." This is not said to minimize those issues. Not at all. It is said instead to underscore the point that the concern subtly reflects that we recognize how important individual teachers are.

So, let's start there. As a society we do seem to recognize (if too often subconsciously) the importance of the teacher in the educational system. This recognition makes the way we collectively allow ourselves to treat them even more puzzling and alarming. We recognize, as the quote goes, that "teaching is the one profession that creates all the other professions." The teacher is *that* important to society. And yet:

- Our teachers are (sometimes terribly) underpaid in relation to the rest of society. We pay our plumbers more.
- Teachers usually have the least power over making general changes that matter for student learning in the classroom, even though they are the ones who are IN the classroom.
- Teachers are often scapegoats for all the wrong reasons, and often for things and policies over which they have no control.
- Teachers often must purchase [even after a minimum allowance] many, if not most, of the supplies they want and need for their classrooms.

- Teachers are constantly burdened with excessively large classrooms. This is highly counterproductive in two dangerous ways: 1) It robs the teacher of extra time that could be given to a student who needs it, or to notice individual student problems when they are small. 2) Teachers do not have time to think, reflect, adjust, or even rest during a day, and sometimes a whole school year. (See ***Making Progress in Neutral*** in Section 7.)
- Teachers are given too little time to visit with colleagues, test fresh ideas, discuss strategies, or share observations. And worse, they are often burdened with so many other "duties" besides teaching.
- Teachers are rarely given time (let alone funding) to attend professional conferences to help them gain new insights into their teaching and their subjects.

Almost certainly, none of these conditions are intentionally or maliciously designed to have this collective effect. Most are the result of funding issues and have developed slowly over time. As such they are not quickly correctable, even when we recognize them.

It is sad, but true. Being a consistently good teacher—day in and day out, year in and year out—may be one of the most difficult jobs in our society. And we continue to make it more difficult.

We desire and demand excellent teachers, and yet our collective actions over time have made it harder to attract them. Have we been penny-wise and dollar-foolish? Is it any wonder that fewer good candidates are choosing teaching, and those that do are burning out more often and more quickly than ever before?

Surely, we can help somehow. Can you simply thank a local teacher? Can you donate money to a school or a classroom, or start a GoFundMe campaign? Can you buy some supplies for a teacher? Can you volunteer your time to help in a classroom? Can you petition your school board to help? Can you start somewhere? If we don't start now, when will we start?

DEAR SANTA

Dear Santa: My name is N.E. Teacher, and I teach in the N.E. School District here in Missouri. I'm sending along a list of things I'd like this Christmas. Heck, it's not just me, though—any teacher would love these things as much as N.E. Teacher.

I realize this list is late, and that these things aren't as tangible as the toys you were just carrying, but look on the bright side: You won't have to lug these things down some chimney to deliver them personally. And, it's not urgent. (I've been without these things so long). The next week or two would be fine.

Where do we start? Truth be told, Santa, what's probably at the top of most of our lists is some public perspective, awareness, and support. Could you help arrange those? Honestly, if we had some more public awareness and perspective, the support would follow automatically.

Here's just one perspective example and I'll try to bring it close to home. Forgive me if I have a wrong impression of your job (folks misunderstand ours, too), but by the time you read this, you'll likely be back at the North Pole, recovering from your exhausting Christmas Eve gig, with Mrs. Claus, and looking forward to a good long break.

Instead, let's imagine that you suddenly have 30 new elf trainees. They're considerably younger, have a lot more energy, and sometimes a lot less "on task" motivation and shorter attention spans. Mix in lots of different skin colors, cultures, and even languages, a good dose of poverty in some of them, and often one- (and no-) parent families. And, oh yes: they have no prior experience.

Then imagine that, starting next week, you have them constantly for 8 hours a day or so, whether they want to be there or not—you can't pick them yourself, or fire them, you know. Your task is to teach them all the important skills they need to be an elf. By the way, you don't have long, as you have a job to do next year, so you need to get them performing up to snuff soon (there'll be a standardized test), and get them prepared for your annual encounters with the real world.

We teachers have all these restraints daily for a full school year, every year. I don't mean to brag, but we still manage to perform miracles at an incredible rate. And with lots of other things trying to distract us, as well.

Hey, I'm not complaining, Santa. I love my job as much as you do! I seldom have time to write a letter to anyone, so perhaps I'm babbling a little. There's so much more I could say, but I have so little time/space, perhaps I should wrap this up.

I'm sure you can see how just a dose of perspective and awareness would almost automatically lead to more support for the work we do. And that support would manifest itself in SO many wonderful ways, and in so many arenas, all the way from our classrooms and local communities to Jefferson City to Washington. (And, we're not trying to do your job, Santa, but there are some folks in both those places that deserve a lump of coal in their stockings.)

Can you help us?

Thanks for your time, Santa. I/we may write more later. In the meantime, we'll leave the light on (and the cookies out) for you. Good luck with those elves!

ARE OUR STUDENTS AFRAID TO THINK?

The comment was so ordinary that I'm surprised it stuck with me.

This was years ago. I had just finished a two-hour workshop on problem-solving with a group of high school upperclassmen in rather difficult surroundings—a high school auditorium. It had been hard to do some of the things I usually did, yet it had gone well—better than I would have expected under the circumstances.

In fact, once the students had gotten rolling, we all had a good time and they had exhibited some really neat ideas and approaches to some of the fun problems. I was impressed with some of their thinking and I mentioned this to the teacher as the auditorium cleared and we were cleaning up.

I forget the exact words, but she smiled and jokingly said something like "Yeah, heaven forbid we get them thinking. They're usually so afraid to think." It was purely an innocent joke, but somehow the words "afraid to think" stuck with me, especially since these students had just shown they were clearly not afraid to think.

I suspect it's something we've all heard or said ourselves: "These students today are afraid to think." There's a paradox here. I honestly don't think students are afraid to think. Do you? And yet, it seems, they *don't* often exhibit good thinking skills, especially in classrooms. Why is that?

I thought about that situation, that comment and that paradox for years. I even developed a talk built around it. I think there are easily a half-dozen important and related reasons why students don't show their thinking skills more often. However, at the risk of rash over-simplification, I'd like to focus on just one.

I'm not sure that students are so much afraid to think. I think, among other things, that they're afraid to be wrong.

Think about it. Being wrong, after all, is what the system has taught them is so highly penalized. From very early on, they learn that what the system values is not so much their thinking, but whether their answer is right. One of the tragic ironies here is that it's some of our most talented students—the ones who should be doing the most thinking—that are also the ones that can least afford to be wrong. From tests to grades to ACT/SAT tests, there is nothing more important than right answers and good scores as they prepare for college entrance and beyond.

Okay, please note. Of course, I'm not saying that right answers are bad. I'm just afraid that we've subtly and unintentionally created an atmosphere that is counterproductive to what we're trying hardest to do, namely getting our kids to think. We've somehow got to remind our students (and ourselves) that getting wrong answers while learning to think is like falling off a bicycle while learning to ride. It's part of the process.

Indeed, the story goes that a successful businessman was asked the secret of his success. He responded, "Two words: Good decisions." When asked how he learned to make good decisions, he responded, "Two words: Bad decisions."

It's a strange, marvelous, and sometimes scary process, this learning how to think. But our students are not afraid to do it. Naturally, this is *much* easier said than done, but let's give our schools and our teachers permission to re-create an atmosphere where thinking skills can flourish.

COVERING MATERIAL OR PROMOTING LEARNING?

For years, leading up to our respective retirements, I co-directed several state-wide professional development projects with a math education colleague from University of Central Missouri, Dr. Terry Goodman. I enjoyed working with Terry for many reasons, among them our similar philosophies (and senses of humor). But another big reason was his good perspective on (math-) educational issues.

I'll always remember one comment Terry made in one of the first programs we were leading together. We were on a mid-day break and visiting informally with several classroom teachers. They were discussing the yearly problem of feeling the need and/or the pressure to "get everything covered" in their classes.

In a perfectly casual manner, with just the right touch of humor, Terry said, "I quit worrying about what I was covering in my classes and started worrying about what my students were covering."

I always liked that statement because it seemed like a clever way to frame what seemed like an obvious fact: "Covering material" doesn't always equate to "student learning," and it takes a good teacher to know when those two are happening together and when they're not.

Knowing when classroom material is being learned in an authentic fashion is one of the toughest jobs any teacher has, and it can be severely hampered by getting forced into the "covering material" trap.

This trap manifests itself in several ways, one of which we've already mentioned: the pressure, usually from some outside source, to "get everything covered" before a certain deadline.

Another way this trap can show up is even more dangerous, because it can seem like a promising idea at first. There are school systems and/or institutions where it is required that all sections of a given class be "covering" exactly the same material at the same time in each class.

On the surface, this seems like it might ensure some consistency of "coverage" across classes. In reality, there are few things that handicap a teacher's ability to do his/her job more than an ill-conceived requirement like this.

I experienced this requirement first-hand once, in a community college course I taught for a semester. It was neither fun nor productive. Inevitably, there would be an occasional class session where it was clear that the material for that class hadn't quite "sunk in" yet with these students. Every teaching instinct I had told me to back off a little and revisit the topic a bit more in the next class. To do otherwise was simply not fair to the students.

And yet, that naturally took time away from what I was supposed to "cover" in the next class. Once you're in such a vicious cycle, it's very hard to extract yourself. The dilemma then becomes whether to honor your employer's wishes and cheat your students, or vice-versa. Not much saps the joys of teaching (or learning) more quickly, and everyone suffers.

The irony (or perhaps the crime?) here, is that in each of these "covering the material" traps, the power is literally taken away from the single most important person in the equation, namely the teacher. We can't continue to hamstring teachers trying to do their jobs, and then try to criticize them if things don't work the way we want.

If there is any good news here it is that there is an escape hatch: Let's allow (and trust) teachers to focus on "authentic student learning" and not confuse that with "covering material."

ASSESSMENT–VERSE 17?

What do we want our students to know, and how do we know when they've learned it? The first question deals roughly with matters pertaining to curriculum. The second question deals with the always-tricky question of assessment.

We've discussed both these questions before, but let's look at yet another aspect of assessment that is both subtle and tricky. Sometimes we can look around and discover that we're unintentionally assessing something other than what we claim we want.

During graduate school, I took a History of Mathematics course that I'll always remember. I liked the course more than I thought I would, and I think I learned a lot as well. But the overriding memory that remains from that class is that my classmates and I didn't really get to show what we had learned.

The professor had a habit of giving overly long exams. Rarely did anyone finish an entire exam, and many of us left each test frustrated that we didn't get to answer questions we knew. I recall scrawling "I know this!" on the still-blank last page of the exam as I was forced to turn it in. I'm sure my protestations did little good (other than therapeutically), because the professor acknowledged that long exams were his goal. I believe it was his way to separate the wheat from the chaff, so to speak.

I'm not sure I could have verbalized it back then, but I now realize that our frustrations were probably at least somewhat appropriate. The professor was not necessarily assessing our knowledge of the material as much as he was also measuring our ability to work quickly, an entirely different thing.

This vivid memory helped me in the years following, as I began to give, rather than take, exams. I'm not sure I always succeeded (assessment is tough!), and doubt if my students always agreed, but in the back of my mind, I tried to make what I was assessing match what I had wanted them to know.

In today's world of rapid and convenient technological access, we are often guilty of falling into the same trap. Raw information (dates, places, names, etc.) and various skills (times tables, square roots, etc.) are no longer as important as they once were. Google can find information, if we need it, and calculators can take a square root, if we need it, and both can be done almost instantly. It's no longer *as important* to assess/evaluate those things, as they are now readily available in our everyday lives.

At the same time, this situation provides a golden opportunity to re-ask the first question above. Yes, we've played this tune before, but instead of emphasizing when/where D-Day took place, e.g., we can discuss things like how D-Day affected the outcome of WWII, and therefore changed our lives. Instead of overly emphasizing times tables, we can better help students learn to use arithmetic (yes, with calculators) to tackle and solve important problems.

The catch is, of course, that it's *not* easy to evaluate those things. It's much easier to test for information and skills (especially on standardized tests). Unfortunately, simple paper/pencil testing isn't always enough for the things we want students to learn.

We must not fall into this assessment trap. We must take the time to ask what we want students to know, and then struggle together to find ways to authentically evaluate when they are making progress.

FLYING, LEARNING, AND TEACHING

I've always had a love/hate relationship with flying.

On the one hand, of course, flying certainly is time-efficient for getting to destinations. And it's often a beautiful, awe-inspiring view from up there, isn't it?

On the other hand, I've always been one of those "flying doubters" who, despite my faith in all things scientific, isn't sure how those huge things get off the ground. I can't always make a whole flight without seventeen irrational fears invading my thoughts. Looking down from a plane's window is a curious mixture of majesty and discomfort for me.

The love and the hate butted heads in my thirties. I had read several of Richard Bach's books. I was captured by his love of flying and that sky which is always perfect. (I still love those books.) I think that was the motivating factor, but for whatever reason, I decided to take flying lessons.

My flight instructor was a great guy and an accomplished pilot. He was congenial, knew his stuff, and I liked him. But I had a tough time learning from him. Perhaps it was his style, perhaps it was my uncertainty. It doesn't matter. What would happen is that we'd be up there on the downwind leg (before turning twice to land), and he'd start rattling off a list of "things to do." "Okay, bring the speed down, trim the flaps, prepare to turn," and several other instructions which rapidly blended together into one blur of sound. I'd be doing one thing and miss two of the instructions. Some students take to flying instantly, but I wasn't one of them.

It was then that I gained an appreciation of how hard it must be to teach flying to folks that don't "take to it" instantly. And it was then that I gained a fresher perspective of the teaching/learning dichotomy.

My perspective about learning came from the fact that, if I was going to do this, I wanted to by-gosh learn to fly, and not just "pass the test." I wanted to know what I was doing. I wouldn't have time to encounter a situation for which I would have to think "Ack, what was I supposed to do here?" (This came perilously close to happening once.)

And my new perspectives about teaching came, of course, from the fact that I taught a subject which, like flying, not everyone quickly takes to. (Did anyone compare my rattling-off-instructions description to their perceptions of their math classrooms?)

How many of us still believe that teaching consists of providing a "here's what to do" list, and that all students successfully and naturally learn that way? And, even if they "pass the test," do they by-gosh know?

Isn't it interesting then, that in our system—at any level—a teacher routinely enters a classroom with some students who instantly take to the subject(s), and handfuls of others who don't. Then, we and the system expect all those students to be equally challenged and to all end up at some "proficiency level" at the same time. And we become worried if they don't. This is especially true in my discipline for which the extremes of "drawn to" and "repelled by" are as varied as those of flying. It's a huge challenge that we don't always address too creatively. Several topics are still begging for more attention here, but I must close.

I eventually learned to fly. I survived all my solo hours and cross-country trips, though I never finished the paperwork for the license. Overall, it was a good experience, because I learned lessons beyond "flying." My experiences in the cockpit provided valuable insights that followed me into the classroom.

THE TALE OF THE MIDDLE SCHOOL ASSIGNMENT

One of my favorite student teaching/learning activities occurred each Fall in a Capstone class for future middle school teachers. It was usually fun for all, and always enlightening for my soon-to-be teachers—and often for me, as well.

Early in the semester, I'd divide the class into three to five small groups. I'd usually provide three varied topics, from which they were to pick one to construct an assignment for a hypothetical class they would teach. They were to work in a group, create one assignment as if they were passing it out in a class, and have this ready by the next class period.

I might note that it was fun to wander around the room and listen to some of the conversations as they began to work on these. It was a good exercise in viewing what they (and others in their group) thought the assignment should include and look like. There were some interesting dynamics there, as they experienced each other's ideas.

Usually, one of the three topic choices was "non-mathy," asking students to pick a favorite mathematician, learn generally about them, and then share in a short paper. Interestingly, this was often the one the future teachers would pick. For this reason, as well as to make the thoughts more generally applicable, I'll focus on that one.

For each assignment, I would later prepare two "middle school" student responses. In each case, the response of "Sally Square" would be as clearly excellent in content as I could make it, but there would be minor instructions not followed. If the requirement was to double space, Sally might forget and single space. If a three-page paper was assigned, Sally might not stop until the fourth page, rather than cut material. You get the idea.

The other response was from "Tommy Triangle." Tommy always followed the letter of the law perfectly, but his was clearly a typical squeak-by submission. It was often sloppily written, perhaps had grammar mistakes, and had obviously been "thrown together" at the last minute.

In the following class, the students were asked to re-group and give each paper a grade or score. All groups always agreed that Sally's assignment was "better" and that she had learned more, but they were often in a quandary about how to score the separate papers. All of them were naturally disappointed in Sally for "not following instructions"—who can blame them?—and they were often astonished to find themselves giving the papers similar grades.

When this happened, their first instinct was to fix their original assignment by establishing more parameters.* Usually they quickly realized that could only make the possible predicaments even worse. I used to gently mention to them that in these cases, sometimes less is more, especially when the goal is assessing learning. I told them of a special middle school teacher I knew who used to add "turn in something your parents would be proud of." This often succeeded better than any rubric.

There was no need to grade this final score-the-papers activity—there were no right answers, after all—but it almost always engendered some great discussions, insights, and reflection.

Primarily, it allowed the students to experience for themselves some truths about assessment that they might not have readily accepted from a "stuffy college prof." 1) Assessing authentic learning is rarely easy, even in a math class. 2) Good assessment must first involve knowing what you want them to know. 3) It's easy to fall into the trap of assessing (or over-assessing) something else, if you're not careful.

* *Author's note: Interestingly, as this is written, there exists a similar, and controversial, situation in the world of sports, namely NFL football. The NFL has gone to extremes to define what a "catch" is, and it has become so complicated that many are beginning to call for an abandoning of the increasingly complex decision, and a return to "less is more." One often hears some of them saying, "We know what a catch is... ."*

SECTION 6
Beyond the Classroom: Administration and Support

INTRODUCTION

The individual classroom, in all its various manifestations, continues to be one of the key elements in the entire educational process, especially in the public-school arena. But the ultimate success in those classrooms depends heavily on factors outside the classroom. One of these factors includes administrative support. Another is public support and trust in general. These two factors go hand-in-hand, and they are both crucial.

A teacher's job is made much easier with valuable administrative support. And ideally, that support exists to make the job of teachers easier by allowing them to promote learning more effectively. Running and managing successful school districts are difficult jobs.

Things get complicated when "the administration" also has authority over things that happen in the classroom. This situation is usually a practical necessity for things to run smoothly, but it can naturally create areas where perspectives are different, and outcomes are sometimes conflicted.

Add to this the crucial component of public support and understanding for its schools (all of which include both teachers and administration), and lots of interesting issues arise. Selections in this section, then, address one or more of these overlapping areas.

TRUST, RESPECT, AND EMPOWERMENT CYCLES

It's an irony that has struck me for decades now, having seen and experienced it so often. Call me naïve, but wouldn't one think, on the surface at least, that the two institutions most likely to practice tolerance, open-mindedness, mutual respect, and teamwork would be churches and educational institutions?

Why is it then that these two influential role models are often the last places we find these admirable types of things happening? It can be disheartening as well as ironic.

My purview is not religion, of course, so let's focus on education. I think higher education can be the worst, but let's focus briefly on the public-school systems, where the trust/respect factor is an essential element of a well-functioning district.

In an earlier column (see ***Chain of Influence***, Section 3), I wrote about what I called the "chain of influence" (or empowerment). On the one hand, we have teachers who answer to administrators who answer to a school board, which answers to the public that elects them. On the other hand, if the system is to work effectively, once the public has elected a school board, they must somehow trust and empower that school board. That board in turn must hire, trust, and empower the administrators, who must hire, trust, and empower the teachers (who are obviously hired to teach and empower the students, the theoretical focus of it all).

It's a two-way ladder where the players on the bottom of the "authority" ladder, namely teachers, are at the top of the influence/empowerment ladder, as they most directly affect the learning in the classroom. It's almost a necessity of the structure, but it can be a tricky situation, requiring lots of mutual trust to insure the best results.

Here's the deal: Find a model school system where exciting things are happening, especially in the classrooms, and you can pretty much guarantee that you'll find respect, trust, and support flowing (fairly) freely back and forth among all these groups. Teacher empowerment is a powerful tool. Conversely, find a system where this flow is jammed in several spots and you can guarantee the big picture results are not that good.

Here's a quick thought exercise: Pick the school system of your choice, and then pick any two of the following groups: parents (citizens), teachers, administrators, school boards. Then answer the following question, after inserting your two groups in any order: Does Group A support, trust, and respect Group B (and vice versa) most of the time? My guess is that, too often at least, the answer is "no" in at least one direction.

Is it any wonder then that our American school systems (and therefore our students) are in the collective trouble they often seem to be in?

There is no "quick fix" solution to this possible trust/support/respect problem. The fix, when needed, is very hard. But, perhaps just identifying and isolating one part of the problem may be a good first step.

All, of us, no matter which and/or how many of the groups we are in, suffer when the trust, support, and empowerment cycles start to break down. The quicker we all work to re-establish those systems *together*, both locally and nationally, the better we will all be. The quality of our children's education depends on it.

STEWARDSHIP AND PRIORITIES

I ran into a former student not long ago, working in a large chain store, and had an opportunity to visit briefly. Running into former students is almost always a fun experience, and that was the case this time—at least until she told the story below.

We'll call this former student Mary. In class, Mary was a very good student, asked good questions, had good insight, and was open to innovative ideas in order to improve her growth. I could tell from the start that she was also going to be a good teacher. Indeed, after graduation, Mary went on to teach middle school math at a local public school. She loved teaching, and by all accounts, did an excellent job.

Then, as is often the case, life interrupted for a while. Both her in-laws became ill and she stepped out of the classroom temporarily to help with their care. After both passed away a few years later, she renewed her desire to get back in the classroom. Then, another bout with life, her husband unexpectedly died at forty-nine.

Very shortly afterwards, a school district called to ask about her availability, and this was a lift she needed. She was ready to go back.

Alas, the plot thickens. She ended up interviewing with three different school districts. All three were impressed. All three told her she was their top candidate. Then all three school boards refused to approve the appointment because she was a veteran teacher and would cost them too much money.

Incidentally, Mary told all three districts she'd be willing to go back to entry level pay. That's how much she loves teaching, but the state of Missouri won't allow that much "step down." (Apparently, they are worried about veteran teachers replacing new teachers.)

I won't question Missouri's policy. Perhaps this is a case of unintended consequences. On the surface, I won't even quibble too much with a school wanting to conserve funds. That can be considered good stewardship.

And yet, here sits this marvelous, experienced, and talented teacher, working a desk job in a retail store, because she can't get hired. She has *too much* experience. What's wrong with this picture?

Okay, for the sake of balance, it is certainly true that this is Mary's story, and I have not chosen to "fact-check" it for accuracy. In my mind, knowing Mary, there is no need to, but just to be clear, I'm going to assume the situation is essentially accurate.

So, let's return to the "good stewardship" possibility. Is stewardship only about money? Wouldn't long-range good stewardship dictate that the best possible teacher be available in every classroom? Aren't decisions like this penny-wise and dollar-foolish? Are we willing to hire a less-qualified teacher (even one with great promise) for the sole purpose of saving money? What will we then achieve? Is a "good school district" one that saves money at every turn or one that hires the best possible teachers?

I'm intrigued by this irony. If life had not interrupted Mary's teaching tenure, my guess is she would still be at her original position, with even more experience, with an even higher salary. And, I suspect, a gladly renewed contract every year. So, where's the difference? Are we soon to be releasing excellent teachers with experience because it costs too much to keep them?

I understand, and sympathize with, the fact that school boards are faced with tough decisions every day. And I applaud their challenging work and dedication. But this situation baffles me. Where are our priorities?

PONDERING A CLASS SIZE OF "ONLY" 20

As seen in Section 2, many of my own personal insights about education and teaching have often arrived in interesting contexts. In one of those Section 2 pieces (***Learning About Learning***), I shared the case of the insight I received in mid-sentence in the middle of a calculus class early in my career. The anecdote here wasn't quite that dramatic on a personal level, but it was probably funnier, at least to an outsider, and it had an influence on my thinking and understanding.

It was the early '90s. I was between stints at institutions of higher learning and was working with community education in my home town. For one month, on Saturday mornings, we ran a class called "Fun with Math" for youngsters. We had nine first and second graders in the class. It was the first time I'd gotten to work directly with that age level, and it was a fun and instructive experience for me. The kids were eager, curious, and had not yet fallen prey to the seductive "one right answer/method" syndrome, so they were anxious to learn. Their behaviors were refreshing and enlightening.

On the third Saturday it seemed that the nine students *and* the cloudburst of a thunderstorm arrived at precisely the same moment. I suspect it was the weather, for suddenly it seemed that each youngster had just received their first introduction to caffeine earlier that morning!

I think the humor that hour would have come from watching me trying to deal with those extra doses of energy and activity. Remember there were only nine students—*and* one of the youngster's mothers (who was also a math teacher) was helping me. We maintained both control and a small amount of decorum, perhaps barely, but there were a couple of brief moments when I remembered why I was glad our own two children were almost grown.

The aha-type insight arrived sometime during that hour and continued after the class, and it was sobering. As I thought about the marvelous energy, eagerness, and curiosity of those youngsters, I also thought about how I was curbing those qualities to maintain some semblance of order and sanity.

And as I pondered, I instantly thought of how our school systems routinely assign at least two (and often three) times as many of these same children to our classroom teachers. And I was sad—for both the teachers and the children.

My younger sister is probably one of the three best kindergarten teachers in the entire St. Louis metropolitan area. We talk often on educational matters and I've been in her classroom. I constantly marvel at what she can do with youngsters that I could never accomplish at that level. She has often told me, sometimes directly, sometimes not so directly, of the devastating effect that a minor increase in class size can have on what she can accomplish, not to mention the energy she has with which to accomplish it.

We know the realities. Like it or not, class sizes are now more of an economic, rather than educational factor, regardless of schools' visions. What district could get a bond passed to cut class size in half (and thereby adding teachers and buildings), even if they were passionate about it?

But I've seen the educational (and human) price we're paying for classes of "only 20," and I can't help wondering, "What if...?"

MAKING PROGRESS IN NEUTRAL

Decades ago, a graduate dean instructing one of my doctoral classes wrote an interesting piece about the importance of taking time during each work day to think, to reflect, and to ask important questions. I kept that article pinned to my bulletin board, and then, for years after one of my moves, I couldn't even find the bulletin board.

Because those thoughts struck a chord with me, I'm always intrigued whenever I re-encounter similar quotes, like this one by Mortimer Adler, "When I have nothing to do... I sit back in a chair and let my mind relax. I do what I call idling. It's as if the motorcar's running but you haven't got it in gear. You have to allow a certain amount of time in which you are doing nothing in order to have things occur to you, to let your mind think."

This seems like good advice for all of us, no matter our profession. But when it comes to public education, it strikes me as vital. And when it comes to giving our teachers crucial time to think, I fear we are sadly missing the mark, to the detriment of all of us.

Our educational system demands so much from our teachers, especially in the way of time, that we are robbing them of some valuable and crucial time to just plain think and reflect, especially during the school day. And this robbery is not just hurting them, it's hurting their students in turn, and therefore, all of us in society.

Among other things, I'm thinking especially of the non-teaching duties we dump on teachers. Attendance, money collection, passes, grade recording, and countless other administrative details in the classroom. Hall duty, lunch room duty, ballgame duties, other duties outside the classroom and even outside the school day. All this and usually—for the elementary teacher anyway, who gets no "planning period"—not a single minute in a day to spare, let alone reflect or plan! (Or even rejuvenate, for heaven's sake— when was the last time you dealt with two dozen or more youngsters for seven hours?)

It is a daily schedule and list of demands that require incredible energy, not to mention excellent planning and organization. With little, if any, time to adapt, to adjust, or to think.

No one can deny the need for the various non-teaching duties and details mentioned above. And, as always, I'm not sure that there is an immediate (or at least an immediate inexpensive) solution to the dilemma.

Still I wonder. Do we really want our teachers, our most valuable resources, doing all these things? We barely give our teachers time enough to "teach," let alone give them any time to put their minds in neutral, to think and reflect, especially when thinking is so crucial: thinking about how the last class went, about how to get a concept across better, about a new assessment technique, let alone about how to communicate an interest in Johnny or a concern for a home problem of Sarah's.

All these "neutral moments" are so crucial to good teaching and learning, and to good, safe learning environments (and to the sanity of the teachers), and yet we continue to rob our teachers of them by stealing—or not providing—this crucial time to think.

ISN'T TEACHING A HIRING PRIORITY?

Having worked with future teachers, I've written a lot of references for teaching jobs. I've recently had occasion to respond to a couple more. As a result, I've noticed (again) a fascinating, puzzling, and highly frustrating phenomenon.

The two most recent recommendations were for different schools of different sizes in different areas of the state, yet they were nearly 100% identical. In and of itself, this is not too surprising. It seems that most schools these days "outsource" their recruiting/hiring services to larger companies, who prepare the online reference forms (for approval?) and probably tabulate the responses as well. These two schools used the same service, I'm sure.

Whether school districts' decisions to use these services is penny-wise and dollar-foolish is probably an excellent question, and I'd love to explore it. But that will have to be left for another time.

What I found both fascinating and frustrating is the nature of the questions themselves. On the surface, the questions look typical. I'm reluctant to take space to list them all, but it is pertinent, so....

After the usual "relationship to applicant" data, I was asked to rate this student on attendance, dependability, willingness to assume responsibility, ability to follow instructions and respond to supervision, and both quality and quantity of work. (I'm still a little puzzled over those last two.)

Then I was given the chance to list the person's strong points and areas that might need improvement. I was also asked if I would "rehire" the person (as if a college professor could rehire a student!). These, by the way, were "required" questions.

These are fine traits, of course, and most are probably desirable qualities in a teacher. On the other hand, have you noticed anything interesting yet?

Here we have a recommendation form for a *future teacher*, and *not one* of the questions relates to *anything* that deals with the candidate's ability to teach. Or relate to students. Or inspire students. Or help struggling students. Or dozens of other related traits I believe we'd want a good teacher to have. There's not even a question about the applicant's knowledge of the subject matter!

How can this be? Why would a district—let alone *several* districts—approve this type of form? How can a district even *appear* to be uninterested in an applicant's ability to teach, as hard as that may be to determine? It appears that good teachers are hired *in spite of* (and not because of) these reference forms.* Do we want to resort to luck?

Are we *really* more interested in a future teacher's ability to "play well with others" and accept supervision, than we are in his/her ability to help students learn? In fact, there is a convincing argument that the *best* teachers tend to be pot-stirrers, but I'll bypass that for the time being.

There's another huge irony here. It's very subtle, and I don't particularly like to bring it up with all its complexities. I've written before about how difficult it is to truly evaluate teachers and the art of teaching. Nonetheless, in an era when everyone (mistakenly, in my opinion) seems determined to judge teachers on the "performance" of their students, how can we appear to be sending reference forms that don't even *try* to find out how well candidates might do with students?

I'm aware that there is a lot more going on here than meets the eye. Often these forms are used to "screen" applicants to decide who to interview, where I'm sure (?) that the ability to teach enters the hiring equation. But do we even want to be *screening* on just these qualities? Shouldn't the *hiring* process of good teachers be as important as their evaluation process?

* *For more insights revealed by these forms, see the next selection* ***Was Your Favorite Teacher Punctual?***

WAS YOUR FAVORITE TEACHER PUNCTUAL?

Think back to your days in school, at any level. Can you name your favorite teachers? It could be kindergarten, it could be post-graduate advisor. Pick your three favorite. My guess is that most of us can probably do this easily.

Now take another moment to think seriously and identify the reason(s) that each of those teachers made your top three. Can you make a quick brief list?

I've heard these kinds of lists from future teachers for years, so I'll bet some of your reasons included things like, "she believed in me when I didn't," or "he made me think," or "she was hard on me when I needed it the most," or "he inspired me to want to do better and learn more." Did I come close? What were *your* reasons?

I think it's also a safe bet that absolutely none of the reasons included things like, "he was punctual and never missed a day," or "she took direction from the principal well," or "she always turned reports in on time, and never missed a turn at cafeteria duty." Please note: those things may be important, and were probably true in your cases, but they weren't the traits that made those teachers your personal favorites.

Perhaps you've guessed that I'm still thinking about the topic from the last column. I shared about two teacher-recommendation forms (and, since then I've received a third) that asked about many of the qualities in the previous paragraph, but *none* of the qualities in the paragraph before. Apparently, there is a "disconnect" between what we seek when we hire teachers, and what really matters in the classroom.

It gets worse. Go back to your favorite-teachers list. I'm willing to bet that not one of your qualities of great teachers included, "they helped me score well on a standardized test. They really increased my performance." (You *might* have said, "He really knew his stuff and could get me to learn it, too," but you probably didn't connect it to performance on a test. Those *are* different.) Again, this was very possibly true—but, again, it's not what made the teacher one of your favorites.

So, look at our situation: Apparently the qualities we look for when we hire teachers differ from the qualities we look for when we evaluate them, which is bad enough. But it's also a very real possibility that *neither* of those sets of qualities match well with what makes a great teacher.

There are many factors behind this confusion, and it helps to note *just three* of them:

1. What makes teachers great or successful will often vary from student to student. Good teachers must know their students, how they learn, and what motivates them.
2. Partly because of this, evaluating good teaching is one of the toughest jobs in education. Good teaching is *so much more* than increasing performance on skills tests.
3. And partly because of all of this, *predicting* good teaching is equally as (if not more) difficult as/than evaluating them.

So, hiring is obviously not an easy task. Amid all this, our best teachers continue to swim upstream against lots of mixed messages about what's expected of them. These teachers need our support and encouragement. And they can certainly do without our almost constant blame for an educational situation which is often not of their doing. These teachers will go on to be favorites to our children and grandchildren, and perhaps we need to clear up our hiring and evaluating acts—not to mention our perspective—to help them do so.

PAYING FOR PUBLIC EDUCATION

—

Background: This column was written not long after a local school bond issue was defeated. I was surprised, and may have reacted a bit too strongly. Nonetheless, I leave it as is.

"Most people are in favor of change, as long as they can continue to do things the same as they always have." Bill Phillips

Here's my own harsh paraphrase of this quote: "Most people are in favor of maintaining quality in education, as long as they don't have to pay for it."

Okay, that's (purposely) over-stated, but, when the dust settles, what other *general* conclusion can we reach from the news that R-12 voters have defeated the Proposition SPS school bond issue?

I know. I'm not naïve. None of us (including this author) wants more taxes. And yes, some of us (including this author) are on fixed incomes with their own kids no longer using the schools. Our own nest has been empty for years.

These circumstances are givens. They have always been there. On the other hand, this also means that there have always been folks who don't really want increased taxes, along with folks whose nests were empty who *have* voted to pass school bonds for *our* children.

In this case, the school board appeared to have done an extensive job of bringing forth a solid plan to help maintain/improve quality, especially in facilities, in a variety of excellent ways, and the voters said no. How can this happen? If a pipe is leaking, do we say no to the plumber?

This is not only a local problem, of course. Are school bond issues becoming harder to pass in general? What happens nationwide when taxpayers choose *not* to help provide the funding to accomplish needed changes, updates, and improvements in our schools?

Note that no one is saying we should give *any* school board a frequent blank check. It is incumbent on taxpayers—and the boards—to make sure that homework is done properly, and that school boards are accountable. Fair enough. But, when that happens, and reasonable requests are made in timely fashion, how can we say no? Am I naïve, after all?

The conventional wisdom is that you get what you pay for. Implicit in that thought is the fact that you don't get what you don't pay for.

When school boards are forced to eliminate neighborhood schools, do we think they like it? When districts continue to cut various programs in the arts or eliminate staff—personal pet peeves—do we think *the districts* are being shortsighted? Or, upon reflection, are they merely exercising due diligence and reluctantly cutting what *we* won't pay for? Who, then, is shortsighted?

The general public's willingness to help fund *public* education has always been a cornerstone of our educational system. This necessity gets more crucial all the time. It gets *urgent* in this political climate where our new Secretary of Education would—and will—gladly cut every public education penny she possibly can.

We don't have to wait for another bond election. Here's a wild idea: Why not start a GoFundMe (or similar) project for your favorite school building or district? Have a cause in mind, if you must, but if so, visit with authorities first, and work *with* them. Hit the streets and social media, get interested parties on the bandwagon, and make a difference. Then wait three to six months and do it again.

Paying for schools is not an expense. It's a necessary investment. And, for everyone's sake, it *cannot* be neglected.

SECTION 7

Math is not a Four-Letter Word

PART A: PERSPECTIVE

INTRODUCTION

For the most part, I work to keep my education columns as general as possible, but let's face it, mathematics *is* a part of education in general. Further, I notice that missed perspectives are often more easily noticed in mathematics contexts and then translated back to the big picture of general education.

Finally, mathematics is, after all, *my* field. (Research mathematicians would prefer I further specify "mathematics education," but that's a whole other story.) I have spent decades dealing with the issues here, and have some small amount of experience, at least, if not expertise.

Mathematics, like death, gets a bad press. It is far and away the most misunderstood of the "academic" areas, and this is sad for all of us. It is also as hard to precisely define (even for mathematicians) as education itself. This inability to define "mathematics," or even to have a feel for what it is and isn't, leads to some interesting misperceptions and attitudes among the public at large. (We could spend a whole column on the phrase "Do the Math.") Moreover, it translates into some almost tragic practices in mathematics classrooms. This is especially true in our pervasive technological society.

It only makes sense that one of the Sections in the book deals with issues, problems, and perspectives in the field of mathematics, and how we teach it. If you have a stereotypical dislike or perception of (what you call) mathematics, I hope you won't skip this section outright. It's harmless, kind of fun, and you might be glad you waded in. I think this is especially true in this **Part A—Perspective.**

THREE REASONS NOT TO HATE MATH

April is usually celebrated each year as Mathematics Education (or Awareness) Month. I'm guessing you may have missed that? Indeed, you may even question the use of the word "celebrate" in that sentence.

But let's go ahead and celebrate a little anyway. And let's try a unique approach that may be more palatable. Let's look at reasons *not to hate* math. There are several good reasons, but for today, we'll have to settle for three.

Reason 1: Math is not Arithmetic (only). This could come as welcome relief to those of us who spent much of fourth grade learning to divide a three-digit number by a two-digit one (with paper and pencil). Arithmetic is one tool of a mathematician, but it is not what continues to attract him/her to the subject—see #2. A sculptor needs a chisel, but the chisel is not what makes the *artist.* (See also ***A Case of Mistaken Identity***, Section 8.)

Reason 2: Math is one of those rare commodities which is both highly useful and incredibly beautiful. On the one extreme, we have a powerful tool which allows us to put a man on the moon, design computers, and power our instant-information society. On the other, we have a subject whose beauty is often overlooked.

What could be more elegant, say, than being able to *prove* that the collection of prime numbers goes on *forever*? Or that the decimal expansion of the square root of two never (ever) ends or repeats. Both proofs have been around since before the birth of Christ, yet are within the understanding of high school seniors. Such elegant, powerful simplicity.

Reason 3: Math really can help develop curiosity, critical thinking, and problem-solving skills. Give a child a calculator and let him/her enter *any* 6-digit number.

> Then ask how many times it takes to hit the square root key (√) to get a 1 before the decimal point. (Try it yourself!) Then give a high school math student the same exercise and ask *why* it always takes five key-punches, no matter the size of the original (6-digit) number. (The grade schooler will likely ask why, too, but, with enough time and no pressure, the high schooler *may* figure it out, and probably understand square root better, in the process.)

Or, suppose you are told (correctly) that if a couple plans to have a four-child family, their "odds" of getting two boys and two girls are *not* 50/50—they're less. Or, that in a room with more than twenty-five people, the chance that at least two people have the same birthday is *better* than 50/50. (A great "bar bet," by the way.)

Aren't you tempted to want to know why? Or even to say, "prove it!" This is where math works its magic. Not as something to be memorized, but as a tool to solve problems and prove results.

There's a real irony in all the above. The funny/sad thing is that there is one group you really don't have to *sell* math to, kids. Usually, they *enjoy* numbers, they *love* to ask questions, and in general are very creative in tackling problems. Somehow, someway, much of this curiosity, creativity, and enjoyment tend to vanish by about fourth grade. Why is that? Is there any chance the "system" teaches it all out of them?

If any of this strikes your fancy, feel free to follow up by e-mail or contact your local math teacher. It's even okay to come by cover of night. We're used to it.

RIDING BIKES AND LEARNING MATH

Have you ever noticed that we really can't *teach* a child to ride a bicycle? We can give them all the tips we want ahead of time (sometimes too many), but we pretty much have to put them on the bike and let them *learn* the process. No other way around it, really.

Naturally, we usually try to maximize the environment for that learning: training wheels, holding the back of the seat for a while, making sure the learning surface is level, etc. But, when the (bike-tire) rubber hits the road, they pretty much have to learn for themselves, while we cheer them on.

I'm going to say this carefully, because the analogy is not perfect, but when we look at the entire educational process from K-12, isn't it true that a lot more of that process is like learning-to-ride-a-bike than we tend to acknowledge, *especially* with the "bigger picture" skills?

When a child falls off a bike early in the process, we don't give them a C+ in "bike riding." We understand that the falls are a relatively necessary part of the entire process. However, we aren't always that enlightened in the educational process, and my own discipline of mathematics can be the worst.

Mathematics, like death, gets a bad press. One reason is that it's so widely misunderstood. In elementary school, we often work with times tables (ugh), moving decimal points, fractions, etc. But (again, I tread softly) those things aren't really *mathematics.* They're *arithmetic.* Arithmetic is an important tool for mathematics, but it's only one tool, and it's not what mathematics is about. In the same way, hammering and sawing are *very* important skills for the carpenter, but it's not what carpentry or building is about.

(Incidentally, be careful before you quibble and say, "well, carpenters don't have to do much hammering and sawing by hand anymore, so those skills aren't *as* important," because that is the *exact* situation in arithmetic. No one *has* to do times tables and long division by hand anymore—calculators are our "power tools." But doing *mathematics* is perhaps more important than ever.)

To oversimplify a little, mathematics is about using tools to tackle and solve real-world problems, in the same way that carpentry is about using tools to build useful and beautiful things. The tools change, but the big picture doesn't. Indeed, I have said repeatedly in teacher workshops that I'm worried, metaphorically, that we're teaching our children to hammer and saw in isolation (and pass the hammering and sawing standardized tests) and *not* teaching them to build things. We're giving C+ grades to kids in hammering without any connections to, say, table-building, or even letting them *try* to build a table (which is often the best way they learn hammering and sawing, isn't it?)

Which, in the end, brings us back around to learning to ride a bicycle. Learning to tackle and solve real-world problems, using appropriate and available tools, is a life-long process, and, like a bicycle, one learns by practice. And it can be as fun as bicycle riding! And, yes, I actually *do* have my own list of suggestions that help maximize that learning environment, but they must wait for a future piece.

We can't teach future bike riders by teaching pedaling in first grade and steering in second. (And if we tried, they'd think bike-riding is boring.) Again, I'm being general, but it is the same with our future problem solvers. We *must* "put them on the bicycle" earlier (brain teasers, puzzles, strategy games, etc.). And, we must allow them to "skin their knees" in the process.

SELLING HORSES AND SOLVING PROBLEMS

Here's a very old brain teaser: A man buys a horse for $60, sells it for $70, buys it back for $80, and sells it a final time for $90. How much money, if any, does the man make on this series of transactions?

For decades, this has been one of my favorite brain teasers to share with students (of all ages) and future teachers, as it has *at least* three believable "answers," and they often all surface with equal regularity. Good discussions. Try it yourself, if you want. No one is watching (or grading). And then have your answer ready in Part B, where we talk about "the answer" (See *Buying and Selling Good Arguments*).

But this brain teaser—and my wife's reaction to it—created a sequence of events that started out as funny, and ended up as highly enlightening, at least for me. Quick background:

1. For some years now, I've been sending out a free weekly mailing every Monday morning. It always includes one of my photos as an amateur photographer, but it also includes other fun "stuff" from week to week, always designed to brighten Monday mornings. Periodically, I include some fun brain teasers and/or a math tidbit of some kind, and none of my subscribers have thrown up yet, at least that I know of.
2. Before her retirement, my wife Pat was a *highly* successful business woman, but she has the stereotypical dislike of all things mathematical. She definitely thinks math is a four-letter word, and she thinks (incorrectly) that she's "bad at math." Sound familiar?

So. A few months back, I included the brain teaser above in one of my Weekly Photo/Sharings. Mostly to have fun with her, I tried to get her interested in tackling it, but she was having none of it, as usual. For some reason (I certainly wouldn't nag), she finally agreed to *listen* to the brain teaser.

When I read it, she said, "Oh, well, *that's* easy!" She pulled out a piece of scrap paper, grabbed a pencil, and in about 20 seconds had her answer. "I don't even need you to confirm it," she said, "I know it's right." (It was.)

First, I laughed with delight. Then I stared. Who was this woman, and what had she done with my wife? Later, I reflected on what had just happened.

It's clear that Pat had viewed this *not* as a "math problem," but as a "business problem." Now she was in familiar territory. She ate business problems much harder than this for breakfast. She cut through the often-confusing nature of this teaser and didn't blink an eye.

Suddenly the implications for the classroom in terms of having students find and solve problems they are actually *interested* in is more than just academic. It's one of the reasons I'm such an enthusiastic fan of fun brain teasers and math-related puzzles and games. Students of all ages enjoy them, and students are secretly solving problems and doing "math" at the same time.

Take this out a level, and this is really the point, isn't it? After all, outside of a classroom, there's no such thing as a "math problem." *Think about that.* I'll even say it again: In the real world, there's no such thing as a "math problem." There are only "real world" problems that may need math tools to solve them.

And now take it out one last step: Education in the "real world" is so much more than doing/knowing "subjects." We don't learn "subjects," we learn "tools." You don't ever need to diagram a sentence, but you need to write well. You don't need to recite important dates, but you need to know the broad lessons of history. It goes on. Being educated is about applying tools we've learned to navigate life's situations.

But does all this mean my wife now loves math? What do you think?

ALL I REALLY NEEDED TO KNOW...

With apologies to Robert Fulghum for borrowing part of his famous book title.

The famous Greek philosopher Plato had an inscription on the archway above the entrance to his Academy. It read, "Let no one who is ignorant of geometry enter here." He was convinced that the study of mathematics was the best general training for the mind and was therefore necessary for future philosophers and those who would lead his ideal state. Similar arguments flourished well into the mid/late 20th century, and one occasionally hears remnants of them still.

There's a thin line to walk here. While it continues to be true that a general working knowledge of some basic mathematics is increasingly important in everyday life, one would be hard pressed to claim that one must study mathematics to succeed in other areas, or even that it is necessary training for the mind, or for life. Not even I would go quite that far.

At the same time, skills learned while studying the broad fields of mathematics *can* also carry over to other endeavors of life. It's fun to think about them, even if for no other reason than reinforcing the "mathematics is so much more than just arithmetic" theme.

Here then, is my partial list of mathematical skills that also translate well to other areas of life:

1. Problem Solving. Maybe this is obvious, maybe not. But it's true that the techniques one uses to tackle and solve math problems—trial and error, defining conditions, systematic elimination, and so many more—are helpful in tackling life's other day-to-day problems.
2. Trial and Error. This one, perhaps surprisingly, is more important than it looks. Usually it's not the shortest or most elegant path to a solution, but it's often the one that yields the

most understanding. If you're stuck on a problem, sometimes it helps to *try something.*

3. Determination. Trying seven things that don't work before finding one that does helps develop grittiness in life, as well as mathematics.
4. Confidence. And when the one thing you try does work, you gradually begin to believe, and then manifest, that solutions are often a matter of time and determination, not necessarily overwhelming intelligence.
5. Creativity. Determination and confidence often lead to willingness to step back and search for the unexpected approach. And, sometimes, temporarily "breaking the rules" yields surprising results.
6. Paying attention to conditions and details. You can't solve a problem which appears to be asking for "three consecutive integers" (I know, why would you want to?) if you don't notice that it really said, "three consecutive *even* integers." And you won't have much success trying to patch things up after a fight with your significant other, if you bring him/her candy while not remembering they're on a diet.
7. Teamwork. Two heads are better than one. If each head has a different idea about how to solve a problem, it's not only a chance for innovative solutions, but it's also a good chance to reinforce patience and listening skills.
8. Humor and Perspective. If one is really stuck on a problem—and even if not—it's always fun to remember that it's probably not the end of the world. Sometimes a good laugh is not only the shortest route to that perspective, but also the break one needs to see a novel approach.

Skills in mathematics and skills in life... how connected are they? One wonders what Plato would say today.

BALL CAPS AND MATHEMATICS

Back when Dave Barry was still writing a syndicated column, we used to get it in our local paper, and it was always fun to read. Years ago, one of my favorites was entitled, "Young People: Point Your Caps Forward and Study Your Math." You just kinda grin already don't you?

It was a good, fun, light-hearted piece that had me laughing out loud several times. I particularly enjoyed one section about kids wearing their ball caps backward, since it seemed I was reading about our own son, who was sixteen at the time and deliberately wore his cap askew most of the time.

To make a long (and funny) story shorter, Dave noted that a recent survey shows that three of every four ("almost 50 percent!" he says) high school students leave school without an understanding of math. So, his premise was this: "Study your math kids. It's important. And, by the way, turn those ball caps around the way they're supposed to be."

Let's start with the ball caps and work our way into the math. As mentioned, our son, then sixteen, never wore his cap "correctly"—usually it was sideways. He clearly *knew* it was non-functional that way, so I gave up suggesting he looked silly (even if he had been listening!).

I think it was a small statement of sorts—an individuality thing—much like his insistence that his socks not match. Since those were by and large his biggest rebellious "statements," we were mostly ecstatic. The point here is that suggesting that a ball cap is "supposed" to be worn another way was an exercise in futility, so we quickly abandoned the suggesting.

What's the link to math and education? It seems to me that suggesting to students that they study their math (or other subjects) because it's important or because it's "supposed" to help is about as productive as telling them to move their ball caps around, or to take their cod liver oil. (If you're under fifty-five, ask your (grand) parents.)

I'm not suggesting that our kids don't always do well in math because they're rebelling. That would be too easy. But, I think that for the most part, our students often give up on math because it seems boring, sterile, capricious, and certainly unconnected to reality. And that's *our* (mostly teachers, but also parents) fault.

I used to shudder when I heard someone say, "I took algebra in high school and have never used it since." Then it dawned on me that from their perspective, they were right. They left high school and never factored another trinomial after that. We (my fellow math teachers and I) didn't make it clear to them that algebra is much more (and more powerful/beautiful) than factoring trinomials and solving pool-draining problems. We stressed skills, but not the big picture. Math's important foundations are in fact used every day in almost every job. When we can show students *that,* algebra and other forms of math begin to approach actually being fun.

A final word to students. Wear those caps however you feel you must. And don't just "study your math" like you'd "take your vitamins." Ask your math teacher to let you in on the adventure. It can be as much fun as wearing your cap backwards.

THE POETRY OF MATHEMATICS–OR VICE VERSA?

To close this section, here's a little diversion—a quick detour, if you will—from the normal fare in this book. Feel free to have some fun.

It's hard to imagine two more diverse crafts, two undertakings more diametrically opposed in nature, than mathematics and poetry, isn't it? If you love, or even prefer one of these more than the other, raise your hand. Well, we expected *that*, didn't we?

It turns out that's not necessarily the case, and I confess I was surprised—delightfully so, I might add—when I first discovered this little enlightening exercise. (I almost said "quiz," but I was hoping to keep you reading during this selection.)

Here is a list of seventeen quotes from history, from real live (well, maybe not anymore) scholars, many of whom you might know of. Each of these quotes is referring to *either* mathematics *or* poetry (or the human engaging in those arts). It is great fun, not to mention highly enlightening, to try to decide which craft (art?) is being referred to.

Because the authors of the quotes could give away the related craft immediately, the list is presented first without authors. See the next two pages. You'll have more fun (really) if you try to make your selections from that list first. Following the list, you'll find the list of authors of the quotes in the same order. That may (or may not) help some. Yes, yes, the "answers" *are* revealed, but you'll have to visit **Appendix 1** to find them, along with another look at the source of the article.

Unless I miss my guess, almost no one will get all these right. (I was not even close.) Even if you try to "outsmart" things by guessing "opposite." In the end, you'll have had fun, you'll probably be quite surprised, *and* you'll have a new appreciation for the similar ways that both mathematicians and poets view their craft and their explorations of the Universe!

Poetry/Mathematics Quotes–Without Author Attribution

—

Directions: Each of these quotes is missing the word mathematics/ mathematician or the word poetry/poet or a similar variation. Try to insert the word you think make the quote complete.

1. _____ is the art of uniting pleasure with truth.
2. To think is thinkable—that is the _____'s aim.
3. All _____ [is] putting the infinite within the finite.
4. The moving power of _____ invention is not reasoning but imagination.
5. When you read and understand _____, comprehending its reach and formal meanings, then you master chaos a little.
6. _____ practice absolute freedom.
7. I think that one possible definition of our modern culture is that it is one in which nine-tenths of our intellectuals can't read any _____.
8. Do not imagine that _____ is hard and crabbed, and repulsive to common sense. It is merely the etherealization of common sense.
9. The merit of _____, in its wildest forms still consists in its truth; truth conveyed to the understanding, not directly by words, but circuitously by means of imaginative associations which serve as conductors.
10. It is a safe rule to apply that, when a _____ or philosophical author writes with a misty profundity, he is talking nonsense.
11. _____ is a habit.
12. ...in _____ you don't understand things, you just get used to them.
13. _____ are all who love—who feel great truths. And tell them.
14. The _____ is perfect only in so far as he is a perfect being, in so far as he perceives the beauty of truth; only then will his work be thorough, transparent, comprehensive, pure, clear, attractive, and even elegant.

15. ...[In these days] the function of _____ as a game... [looms] larger than its function as a search for truth.

16. A thorough advocate in a just cause, a penetrating _____ facing the starry heavens, both alike bear the semblance of divinity.

17. _____ is getting something right in language.

1. Samuel Johnson
2. Cassius J. Keyser
3. Robert Browning
4. A. DeMorgan
5. Stephen Spender
6. Henry Adams
7. Randall Jarrell
8. Lord Kelvin
9. T. B. Macaulay
10. A. N. Whitehead
11. C. Day-Lewis
12. John von Neumann
13. P. J. Bailey Festus
14. Goethe
15. C. Day-Lewis
16. Goethe
17. Howard Nemerov

Part B: Thinking About Mathematical Thinking

INTRODUCTION

Part of the confusion about mathematics, which often results in its less-than-stellar reputation, is the people not only confuse "mathematics" with the tool of "arithmetic" on one hand, but also with some of its more powerful tools (algebra, geometry, calculus, and so on) on the other. The focus on the "tools" (simple or powerful) obscures the more important process of learning how and when to use those tools to help us tackle and solve problems, which is what mathematics is and is about.

One of the most important parts of learning mathematics is learning to think mathematically, so to speak, with or without tools. One needs plenty of practice in solving problems, and it helps to get a lot of this practice well before many of the more powerful tools are encountered.

If we want to conquer any math-phobia we may have, we must eventually at least *talk about* solving problems in the real world. This problem-solving—or mathematical thinking—is really what math-types do in the long run.

I've tried to select columns for this section that both illustrate parts of this process, *and* still allow us to have fun at the same time. If you're not a typical fan of math, I hope you're still with us from **Part A**, and that you'll consider accompanying us into **Part B**.

BUYING AND SELLING GOOD ARGUMENTS

—

In Part A of this Section, I talked about my wife's surprising reaction to a "Brain Teaser" I'd shared. (See ***Selling Horses and Solving Problems.***)

In that column, I shared the brain teaser, without giving "the answer." I know this was heresy, but it was deliberate. In the first place, I hoped the reader would be curious to know the answer, and perhaps be tempted to tackle it.

But I also had this discussion in mind as well. I mentioned that the brain teaser often produced three common answers, each one of which seems reasonable at first glance. Often, I would use this phenomenon with future teachers to discuss the difference between "good" arguments, and "valid" arguments. I contend these two are not necessarily the same.

Naturally, we want to produce students who can make valid reasoning arguments. In this process, however, we often overlook the value and importance of forming "good" arguments, even if they don't turn out to be "valid." And then learning the difference.

Let's illustrate this. Here's the original **Brain Teaser:** *A man buys a horse for $60, sells it for $70, buys it back for $80, and sells it one last time for $90. Did the man make any money in the transactions, and if so, how much?*

Here are the three most common explanations I'd see, in increasing order of 'money made.' Note that these arguments are somewhat vaguely stated. This is also deliberate, as it becomes part of the point.

> **Solution 1:** The man made $10. He is ahead $10 after he sells the horse the first time, but is back to even after re-purchasing it, then goes back ahead by $10 in the end.
>
> **Solution 2:** He made $20. The man makes $10 on each transaction for a total of $20.
>
> **Solution 3:** He made $30. He originally spends $60 and eventually ends up with $90, for a net profit of $30. All else is distraction.

What do you think? If you're like me, then each of these arguments *sounds* reasonable, and, at the same time, each of them *sounds* somewhat suspicious. But of course, only one argument is "the right answer." How do we know which one?

The fact that each of these explanations sounds reasonable is fine. "Reasonable" should be the first test. (An argument that the man lost money, e.g, isn't too reasonable.) Quite honestly, I'd be **delighted** to have a fourth-grader give any of these three explanations—at first.

The key is what happens next. If you share these arguments with fourth graders (hopefully *they* have produced them themselves), they will know that only one can be "right," and I *absolutely guarantee* they will want to know which one it is. They will be *eager* to know this! A good discussion here does wonders. They are forced (though not unwillingly) to explore reasoning in each argument—which of the "good" arguments turns out to be "right" (valid), and where are the glitches in the other two? *How* can we get to the bottom of this? I've even seen students (encouraged by their teachers) take out play money, and "a horse" and act it out.

By this time, so much "good stuff" has happened to a group of fourth-graders. Without even being able to articulate it, they have appreciated that an argument needs to be good/reasonable first, then examined to see if it's valid. To do this, precision, effective communication, and most importantly, good thinking must be called upon. And they've done some good mathematical thinking in the process. What more can we ask for?

We've said this before, but we don't chastise children for falling off a bike when they're learning. We know that's part of the process. In the same way, as part of the process, we need to allow and encourage good arguments as we move down the road to valid arguments.

P.S. Oh, yes, "*the* Answer!" In a nutshell, the man spent a *total* of \$140 (\$60 + \$80), and received a *total* of \$160, for a net profit of \$20.

WHEN COUNTING IS MORE THAN COUNTING

In the "real world," there are lots of "how many?" questions where it isn't practical to physically *count*: How many license plates are possible in Missouri [and will that be enough to handle the number of cars]? Or, how many telephone area codes are possible [and with the proliferation of cell phones, will we ever run out]? And so on.

Years ago, I had occasion to enter a Springfield restaurant which unintentionally gave me a miniature version of one of those problems and a rich source of classroom discussions, as well.

The restaurant's menu that day had a large and very colorful insert advertising that month's special. The insert pictured nine (9) different options for entrees, and you could "Pick Any Two" for one price." Elsewhere on the insert was an appetizing picture of five different platefuls, and the caption: "Here are 5 of the 81 possible combinations."

I make a confession: My immediate gut-level feeling was "There can't be 81 possible combinations!" Suddenly, my curiosity was jarred awake, and I *had* to check this for myself. I pulled out the proverbial [paper] napkin and pen and started scribbling. I didn't want to have to count. (And, even so, what if I missed an option?) I needed a reliable short cut.

Now, it's true that such a practical shortcut exists, and I satisfied myself of the answer (one way or the other), but for the moment, let's look closer. (As a matter of perspective, it is true that nearly *every* "technique" or "algorithm" to which we are introduced in schools is actually a "shortcut." See the next paragraphs also.)

It turns out that this real-world situation makes a wonderful problem with which to grab students and introduce them to this topic of "counting methods." It also makes a great problem to discuss with teachers as well, for a variety of reasons.

Nine times out of ten, students also *want* to know what the right answer is (and, of course, whether the restaurant chain made a mistake). They are willing to discuss various techniques to find the answer (including counting, of course), make conjectures, share ideas, and other sorts of things that are signs of good problem solving. And, later, when they are introduced to the various "shortcuts," they marvel at them, and more important, they *remember* them. Further, they remember them *as helpful shortcuts,* rather than *formulas-to-be-memorized,* presented to them in isolation.

Other good things happen as well. Often, they begin to ask wonderful questions they don't normally ask. Especially those that deal with *conditions.* The first time a solver asked "could I repeat an entrée? May I choose blackened chicken *and* blackened chicken, for example?", I was ecstatic. Partly because of the question itself, yes, but partly because I had to answer, "I don't know, I didn't ask!" So, depending on a restaurant's *conditions*, there are different *solutions* possible to the question "How many options are there?"

So—and you saw this coming, didn't you?—I will temporarily leave the problem with you to have fun with. What do *you* think? Was the restaurant chain right? Are there indeed 81 possibilities, and more importantly, can you give an *argument* for your own answer? Are there *other* conditions? I'd be delighted to hear any of your thoughts. And we'll summarize and tie up loose ends in a future essay later in this section.

One final reminder. What's going on above *is* very much what mathematics is about. (See Introduction.) More than moving decimal points or factoring polynomials, math is about solving real-world problems. Happy counting.

COUNTING THE "COUNTING PROBLEM" ANSWERS

—

If you've just joined us, our last episode ended while we were still sitting in a restaurant, pondering our situation: We had 9 choices for entrees, from which we could pick any 2 for the daily special. The menu suggested there were 81 combinations from which to choose, and we weren't sure we agreed.

Reader responses to this column (and its set-up) were both fun and varied. And, by the way, noticeably free from the "oh god, I'm supposed to know the answer" feeling often encountered in a classroom. Part of the point, naturally.

When the dust settled, the majority (but not all) of respondents thought the restaurant was wrong. I received proposed solutions of 36, 45, and 48, as well as the restaurant's 81. (Further, more than one of these *could* be right.) Surprisingly, I did *not* receive a submission of 72, which I had expected. See below.

One reader submitted her answer of 36. Here is her (slightly edited) approach:

I used an easier problem of 4 entrees so I could find the pattern. I used ABCD and saw that there were 3 choices for A, 2 for choice B, 1 for C, and no new choices for D. So, with 9 entrees it would be 8+7+6 etc, or 36 different choices.

Did you get the idea? This is a nice approach, and I particularly liked the "used an easier problem" as a first step.

Another reader also arrived at 36, and further noted that if you allow two of the same choice (chicken and chicken, remember?), that would add 9 meals to the total, bringing the total choices to 45.

A reader who is also a teacher volunteered that "problems like this made good homework problems and got family members involved, as well!"

So where did the "81" answer come from, and what's 'wrong' with it? *Maybe* nothing. It appears the menu designer multiplied nine meals times nine (each of nine entrees can be paired with each of the other nine, allowing repeats) to get 81.

This is what often occurs to solvers first. Then, they usually decide *not* to allow repeats (chicken *and* chicken), so they pair each entree with eight others, giving the 72 (9x8) meals I had anticipated I might get. Note that, for my money, this is good reasoning. (See the discussion in the first piece in this section concerning "good" versus "valid.") Where's the hidden flaw in any of these arguments?

When the above happens, the conscientious problem-solver often proudly displays nine columns of eight meals each. One can then often point to a choice (say, AC in column one) and then find its counterpart elsewhere (CA in column three) and ask, "is it a different choice if you turn the plate around?"

The response is usually a broad grin (and a forehead slap?) which indicates the light coming on. At this point, they often add, "Okay, divide 72 in half and you get 36!" Another very good approach. (One submitter above even noted, "I decided chicken and grits is the same as grits and chicken.")

Interestingly, every so often, the (smart-aleck?) response to that plate-turning question is "Yes, it *is* different. I eat my meals from left to right." And, note: *with that* condition, 72 (or 81, allowing repeats) becomes a "right" answer. Ah, conditions!

So, in real life, with these variables, we usually agree on 36 or 45 possibilities.

I must add this fun PS: A teacher once sent me an ad from a St. Louis newspaper featuring a similar display for the same restaurant chain. The ad proclaimed: "Here are 3 of the 54 choices possible." 54? *Where* did the 54 came from? Any thoughts out there?

BOXING GUMBALLS AND SOLVING PROBLEMS

You have 5,000 gumballs. You want to put them in a container for a Christmas present for a young relative. What size container would you need?

Are any of you still reading? Did you bail at the thought/sight of a dreaded math "word problem?" Or maybe you saw that as a fun brain teaser (which it is), which *may* even have intrigued you, but not particularly enough to tackle (especially since you'd have to have more info).

Doesn't matter. For my sister-in-law, it was neither. It was a "real world" problem that needed solving, preferably before gift-wrapping time. My sister-in-law is one sharp cookie, as they say, so I think she already had a plan of attack—maybe even a solution—but she sent the problem to me, hoping for verification. (Math types are always "on the spot" like this.)

Now, granted the solution of this problem does not contribute to world peace (except maybe in one small corner of my sister-in-law's house) or help put a satellite in orbit or reduce the national debt. But the situation *is* "real world" and *does* touch on several points that relate to preparing our students to tackle (and hopefully solve) the variety of problems they will encounter in life. So, begging your indulgence, let's take a brief diversion and discuss it.

FIRST: I suspect you never saw this problem in a text. I think there must be this general feeling that problems in the real world all fit some nice neat category/method that we were taught in school, but have somehow forgotten. (Okay, is this a pool-draining problem, or a "rate-time-distance" problem, and what am I "supposed" to do?)

I'd go so far as to say that this is *rarely* the way it happens. Problems come in all shapes and don't necessarily fit examples in texts. I'll repeat an insight that I shared in a previous column: ***In the real world there is no such thing as a "math problem." There are only "real world" problems that may need math to help solve them.***

So, the bad news is, there is no set approach. But the good news is, there is no set approach. One person's panic is another person's freedom.

SECOND: The solution here does *not* have to be exact. Not at all. We probably only want it to be practical. We don't care if there's a little extra space in the final container (that's why God invented Styrofoam packing bubbles), we just want to reasonably hold them without renting a dump truck. More freedom.

THIRD: We need a little more information, eh? What size are the gumballs? I had a real dilemma when I stated the problem earlier: do I include that data up front or not? Some folks almost literally choke on information in a problem, while others are frustrated if they don't immediately have enough to help them.

Either way, I will tell you now: the gumballs are 0.5 inches in diameter (across). If you get a solution you'd like to share, e-mail me. There's a lot to follow up on here.

One of my favorite maxims with teachers is that the way to spot (and help) inexperienced problem solvers is that they "don't know what to do when they don't know what to do." As we prepare our problem-solvers of the future, they need to know that not knowing what to do—or even where to start—is normal. Once they have some good math-thinking tools in their tool bags, there are always *lots* of ways to start, and eventually solve, a problem, even if it's about gumballs.

GUMBALLS & CALENDAR CUBES—HOW PROBLEMS GET SOLVED

I was/am always *delighted* to receive responses to my columns, *especially* when those responses speak to invitations to send in solutions to problems or brain teasers I have mentioned for one reason or another. (I don't do this often, of course—I don't want to run off readers—but it's been known to happen. And, as you've seen this section is full of them.)

Probably the most—and widest variety of—reader responses and solutions I received came in after the previous column about Gumballs. I hadn't planned to revisit that column, and had to change my mind. Here's a restatement of the problem:

Problem: What size container would be needed to contain 5000 gumballs for a unique Christmas present? The gumballs are each 0.5 inches in diameter (distance across).

I loved the number and the variety of responses to this problem. Almost all of them were different, which was part of the point. Indeed, some of them would "work" in the real world. Other submissions incorporated some *very good* and even elegant mathematics, but wouldn't quite "work," due to an understandable omission. Since all these solutions contribute to a discussion of real-world problem solving, I want to follow up.

Let's start with the solutions that wouldn't "work." At least two of these fell in this category, and they came from an engineer and a former math teacher! *Each* of them used some very powerful mathematics: Using the correct formula, they precisely figured the volume of one gumball, then multiplied by 5,000 and worked from there to get the size of container needed. This would be good, except for one thing. Think about it: The total volume of the 5,000 gumballs is less than the size container needed, since the gumballs don't *pack* neatly. There is a lot of dead space between them!

I guess my own favorite submission (subjective, of course) was from a reader named Michael and his twelve-year old daughter, Lilly. It went like this (emphasis added):

"My 12-year-old daughter and I enjoy these types of math problems, so we put our thoughts together to come up with an answer. Here is our logic.

"Due to the dead space that needs to be included when stacking spheres or gumballs, we took the 0.5-inch diameter of each gumball and cubed it to find the volume of a cube required to hold each gumball. 0.5 x 0.5 x 0.5 = 0.125 cubic inches. We then multiplied that times 5000 gumballs giving us a total cubic inch requirement of 625 for the inside of the container. The cube root of 625 is 8.55 inches, so if the box had the same dimensions on all six sides it would need to be at least 8.55 inches tall, 8.55 inches wide, and 8.55 inches deep."

How marvelous. Not only did Michael and Lilly "nail" the problem *(though, again, several other specific answers are possible)*, they remembered to account for "dead space" between the gumballs (as mentioned, a common Achilles heel for other solutions). And my own personal favorite part: A father and daughter loving and working on such problems together!

By the way, did you note above that I avoided using the word "correct" as I described answers that would "work?" Another thing I like about this "real world" problem is that *several* other specific container shapes/sizes answers are possible—and indeed were received as submissions. Michael and Lilly used 8.55 inches for each side, but a box 9 inches (or more) on each side would be fine, of course. (We're just after practicality.) Further, the box would not even have to be square, of course—a more "rectangular" box of, say 10 x 10 x 8 would be fine, too.

This is an important aspect of problem-solving in the real world. "Mathematically correct" can be a possible hang-up, if different approaches or forms of the answer are involved.*

By the way, I have to mention another "solution" I grinned at. A reader said:

"The solution is to order them from Amazon and they will figure out the size of the container and have it on your doorstep in two days."

It is my own opinion that this "solution" should be honored for its practicality (and its ingenuity). We need those things. On the other hand, if you already have the gumballs (as my sister-in-law did), it's not actually a solution to the problem in question, which also happens a lot in the real world. And besides, from my perspective (which is the minority, I know), it robs you of the joy of figuring it out.

* *For those interested in my own personal version of the solution (similar to, but also different from the one from Michael and Lilly), you may visit this link:* *http://aftermathenterprises.com/nov-15-bts-campbells-bonus-solution/* *, which is repeated in* ***Appendix 1–Related Resources.***

—

This final selection is longer than most, but I want to share responses to one more problem that generated interest. That was the "calendar cubes" problem (first mentioned in a footnote in ***Live and Learn...*** in Section 2). The problem statement, roughly: **Arrange the digits 0 to 9 on two blank wooden cubes in order to be able to represent each date in any given month.** I suspect you've seen such calendars in banks, etc. It's a fun challenge, partly because there's a subtle "trap" or two lying in wait. (In real world problems, those things happen—they are not "tricks" by evil math teachers). **To end up solving it,** there's one subtle thing you must notice first, and then, after that, there's a nice "brick wall" in your way.

My favorite response came from a reader named Dan, proudly claiming to be eighty-one and a lover of all things needing logical approaches. It's my favorite for a couple of reasons, but mainly because it illustrates perfectly the discovery of these "traps" *as well as* the problem solving try-again approach that often is necessary. I'll let our e-mail exchanges illustrate this:

The first response from this reader went this way:

"Larry,

Thanks for the great article. I solved this with the aid of my wife. First I figured that a cube has six sides. Then we had to tell how many duplicates which are 11 and 22. Thank goodness there is no 33rd day. We then realized that there must be 0, 1, and 2 on each cube and the remaining six digits are split up.

I am 81 years old and I still do algebra problems from a book and play chess against the computer. Both are pure logic. Algebra is not really math, only logic. My wife does Sudoku which is also logic, not arithmetic."

Our reader had noted the subtle thing mentioned above: *You need a 0 on each cube!* (You can't put a 0 on one cube and fit all nine other digits on the other.) This is often missed *in the initial stages* of the process.

But Dan had miscounted. After that, there are seven remaining digits to place, not six. (Which creates the brick wall: Seven digits on only six remaining faces. How can we do this?)

When I wrote Dan back, congratulating him, and pointing out the glitch, I get this very quick response:

"I looked at it again and saw that it required one more number than would fit. And then it dawned on me: The 6 could also be used as a 9 if turned upside down!"

Sometimes frustration/mini-failure leads to the eventual solution.

SECTION 8
Technology: Enabling or Crippling?

INTRODUCTION

Few issues in the educational arena seem to generate as much controversy as the issue of appropriate use of technology in classrooms, as we work to prepare our students for their futures. Depending on who has the floor at the given moment, it seems one can hear that technology is the work of the devil himself, or that is a liberating force in authentic learning.

I've always thought we might be asking the wrong question in these debates. Instead of asking "should we use technology?" shouldn't we be asking, "what is it we want out students to be learning?", and then following with "what are the best ways to help them learn that?" If the answer to the second question suggests that technology can help authentic learning in this area, then so be it. (If not, that's fine, too.)

This section follows the one on mathematics, since it seems that the "technology issue" seems to manifest itself more often (or at least with more controversy) in that field, especially as mathematics is sometimes confused with calculating. But the issues involved appear throughout educational arenas, and some of those broader issues are discussed as well.

Author's Note: This essay, by itself, might technically fit better in Section 7A. On the other hand, this column lead to a vehement Letter-to-the-Editor response that was definitely (anti-) technological in nature, and that, in turn, lead to my own follow-up column (which follows this one).

A CASE OF MISTAKEN IDENTITY

Years ago, a cartoon in the *Chronicle of Higher Education* jumped out at me. I ended up using it frequently in workshops with teachers for the rest of my career.

Here's the picture: A traditionally stereotyped, matronly female teacher sits at her traditionally stereotyped desk (in the front center of the room), watching a traditionally stereotyped Johnny do some arithmetic problems on the blackboard (remember those?). She is frustrated and scolds him "Your math skills are horrible! How do you expect to get a job if you can't add and subtract?" Little Johnny answers brightly, "No sweat! I'm going to be a Congressman!"

Many of us will laugh (and/or cry?) at that cartoon, but, in my opinion, the humor is masking a case of mistaken identity. And it's one I worry is still prevalent today.

The mistaken identity is this, in six short words: Math and arithmetic are not identical. We're broached this topic before, but it's always worth another visit from a new angle. The importance and ramifications of this mistaken identity cannot be overstated.

Naturally, of course, *arithmetic* is a part of *mathematics*. But the two subjects are not interchangeable. Any more than punctuation and skillful writing are interchangeable. For countless years, there grew up this impression that mastering six to eight years of paper/pencil arithmetic (that is often timed) is what mathematics is about. Does this impression still linger?

This wasn't such a terrible mistake "back in the day," when higher math necessarily required great deals of calculating, and our jobs in the workplace often required extensive shop-keeping skills without the benefit of a calculator. Knowing one's times-tables was more than handy—it was practically necessary.

Clearly, however the world and the workplace have changed, and drastically. And because of that, the classroom, and the mathematics skills taught there, are necessarily changing too. But not always as fast. Do we really want, for example, to spend much, if any, time anymore on learning times-tables and other purely arithmetic procedures? When was the last time those were used in the workplace, especially where time is money? It's not unlike continuing to spend class time learning to saddle a horse so that one can travel later.

Not only has the world changed, but so have the basic skills needed to survive in it. We need to focus on helping our students learn to tackle and solve problems, using the tools they have at their disposal. This is *not* minimizing the necessity of learning important skills. It is instead to reinforce that nowadays, almost two decades into the technologically oriented 21st century, we need to be sure to know what those skills are.

Perhaps I'm out of date to think this mistaken identity still exists. I hope so. But every time I hear the careless phrase "do the math," I wonder. Math and arithmetic are related, but one is not the other.

So, let's return to Johnny, our aspiring congressman from above. I don't believe Johnny will be an effective congressman without having a good working knowledge of things like statistics, estimation skills, problem solving, interpreting graphs/spreadsheets, handling data, and even number sense. (How many of our politicians really know the difference between a billion and a trillion?)

But I do believe Johnny can learn and use these important skills successfully, whether he remembers (or even learns) his 12-times tables, e.g. (or uses a calculator). And that fact continues to have increasingly pertinent implications for the classrooms of our community and our nation.

THE *REAL* BASIC SKILLS

Long story short: The writer of a Letter to the Editor recently took me to task over the previous column. She strongly disagreed with my opinions that arithmetic skills are no longer ***as*** important as they used to be, and that math and arithmetic are not identical.

Part of her problem seems to be that I was a math educator. (Don't tell my mom. She still thinks I was a saloon bartender.) Apparently, this means I am unprofessional, don't use math in the real world, and disqualifies me from any meaningful discussions about basic skills and how students learn them.

Aside from the letter's tone, I really have no quarrel with the writer. She's entitled to her opinion, after all, and a couple of her views aren't unusual. But I do think some observations about her points are in order:

- This always-present objection about cashiers who can't calculate change quickly is simply not a technology problem. (Nor is being late, phone alarm or not). Some of us can remember such cashiers (and late arrivers) well before calculators and smart phones were invented.
- I will gladly agree that the ability to tackle and solve real world problems is the goal of mathematics. It is partly why I believe we need to spend more time on that skill these days, and less on "adding and subtracting large numbers."
- I would quickly agree that math anxiety can start when students fall behind in arithmetic skills. Part of my point exactly. (Especially timed arithmetic practice! Students begin to think they're "dumb," when they aren't, simply because they can't do something quickly.) Why develop anxiety over "skills" that simply aren't as crucial anymore?
- The writer seems highly enamored with engineers. I would ask her to show me an engineering firm that still calculates with paper and pencil, or "long multiplication and division."

Digging deeper, I believe the subtle crux of the disagreements here lies in the terms "basic skills" and "fluency." The writer makes a big deal that basic skills are needed for "higher math" and problem solving. It certainly sounds reasonable, and I partially agree. So let's stop and ask: "Just what are basic skills in 2017? What were they *ever*?"

Thought Experiment: Consider Student A who can easily and accurately divide 396 by 52 using long division on paper/pencil, but who is stumped by the question "If John donates $396 to a cause in one year, what is his average weekly donation?" Would we say Student A is fluent in division? Which is the "basic skill?" Moreover, if Student B knows that "division" is called for here, does it really matter *how* he gets that answer—mentally, abacus, pencil, or calculator?

This has been our confusion for decades, *even before technology*. Basic skills involve knowing *when* an operation is called for, knowing *how* to apply that operation (with technique of choice), and then knowing how to *interpret* the answer in the real world. Being able to learn and flawlessly perform some rote technique to get an answer isn't a *necessary* basic skill any longer. It used to be our only route to getting those answers, but no more. If we are in trouble, then it is because we have mistakenly focused (even before calculators) on *techniques* rather than *recognition and application of skills* (part of problem solving).

We *all* want our students to be able to efficiently tackle and solve "real world" problems, using whatever tools/techniques are at their disposal. We should continue to focus on that common goal. And in the process, using a calculator to get past the tedious calculations does not hurt—indeed, helps with—the *real* basic skills.

BIG CHIEF TABLETS AND SQUARE ROOTS?

I thought I had my next column written that morning. Then I picked the morning paper out of the driveway, (yes, I still prefer the "paper" paper. I'm old, I guess) and things changed.

An opinion column that morning caught my eye, and then my interest, and then my concern. I worried about its conclusions, and coming from a math educational background, I'd like to discuss it.

The column was entitled *Machines Smarter, Math Scores Still Falling,* and its premise was that math scores "are plummeting," though it does not say where. Locally? State-wide? Nationally? Nor does it provide any evidence of this, though I *do* understand columns have space limitations.

I'm not so sure it's true nationally, but let's not quibble. Let's suppose math scores *are* falling—somewhere, anywhere, or even everywhere. And certainly, our country's performances in math have been cause for concern, especially compared to other countries, at times. This comparison is worrisome whether the actual scores are currently rising *or* falling, by the way.

Here's the rub: The column seems to imply a connection between falling math scores and the existence of faster calculating "devices." Let's examine that.

First, wouldn't it make more sense to make that connection if math scores were *rising?* ("Of course, students are getting more right answers—they've got calculators!") I wouldn't agree with the logic then, either, but I digress.

The column says, "Machines can now work about as fast as our mind, maybe faster," and follows with this question: "So, why hadn't this made math scores skyrocket?" Very ironically, the column *excellently* answers its own question, earlier in the same paragraph: "Technology only makes processes faster, so information can be obtained more quickly." Amen.

Let's think about that. Mathematics is about solving problems (using a tool bag of collected skills), not "gathering information" (though that happens). (See also the perspectives in Section 7B: *Math Thinking*.) This has always been true, regardless of whether we're using slate boards or calculators. Technology itself, faster or not, does *not* solve problems, as the column admits. Faster technology only gets information faster. Mathematics is about *using* that information correctly to find solutions to real-world problems.

If your "tree guy" mistakenly lets a tree fall on your carport, do you blame it on his fancy new chainsaw? (In fact, you might, if he/she hasn't learned to use it correctly.) So, *if* math scores are in fact falling, let's not blame it on faster information-gathering tools. Let's examine why students aren't using that information better.

The column also voiced a complaint that today's students can't take a square root (on a Big Chief tablet—remember those?). I couldn't help but grin broadly. Still, I'm *puzzled* with the implication this is somehow connected to any falling math scores. Where's the connection? Would we denigrate a student who could no longer saddle a horse? If a student knows that a "square root" is needed in a problem (the *key* skill), does it matter whether she calculates it by hand, or saves a lot of time by using a calculator?

I have my doctorate in the field of math, and I readily (and thankfully) confess that I can no longer do the square root calculation. Nonetheless, other than the occasional bouts with approaching senility, I'm not too worried about my math abilities. (Indeed, I say *hooray* to the math teacher who said they hadn't done that in decades. I'd be more worried if they *were* still doing it.)

There are *so many* more issues in play here, and they are important. But if and when math scores are falling, we all do ourselves a disservice if we become sidetracked with spurious connections, rather than seeking authentic solutions.

THE MORE THINGS CHANGE

"One [topic that can help prepare students], is one that is rapidly being phased out by technology. It is the ability for a student to express himself/herself on paper, in the written mode, using thought and creativity."

I'm not sure if the reader/commenter intended to include the words "on paper" or not. Perhaps it was unconsciously added without thinking. But those two words make quite the difference in the context.

On the one hand, I could not agree more that everyone needs to *be able* to express themselves thoughtfully (in any mode), but I wonder if they really need to be able do it *on paper.* Are writing skills "being phased out by technology?"

Having experienced both "paper/pencil" and "digital" writing during school and my career, I now consider myself to be a better writer while using technology than I was before its availability. That may not be a universal opinion, but even that underscores the point: The ability to thoughtfully express oneself is not a function of the medium used to do so.

The opening quote reminded me that "technology" has *always* been changing. It also reminded me of the implications for education amidst times of change (which is to say always).

Long ago, I discovered a fun page of quotes about change that I used to use with teachers. It was titled The More Things Change, and looked like this (with some minor editing):

> **1703:** Students today can't prepare bark to calculate their problems. They depend upon slates. When their slate is dropped and it breaks, they will be unable to write! *Teachers Conference, 1703*
>
> **1815:** Students today depend upon paper too much. They don't know how to write on slate without chalk dust all over themselves. What will they do when they run out of paper? *Principal's Association, 1815*

> **1907**: Students today depend upon ink. They don't know how to use a pen knife to sharpen a pencil. Pen and ink will never replace the pencil. *National Association of Teachers, 1907*
>
> **1929:** Students today depend upon store-bought ink. When they run out of ink they will be unable to write words or ciphers until their next trip to the settlement. A sad commentary on modern education. *The Rural American Teacher, 1929*
>
> **1941:** Students today depend upon expensive fountain pens. They can no longer write with a straight pen and nib (not to mention sharpening their own quills). We parents must not allow them to wallow in such luxury to the detriment of learning how to cope in the real business world, which is not so extravagant. *PTA Gazette, 1941*
>
> **1950:** Ball point pens will be the ruin of education in our country. Students use these devices and then throw them away. The America virtues of thrift and frugality are being discarded. Business and banks will never allow such expensive luxuries! *Federal Teacher, 1950*

There's always the chance (likelihood?) these quotes were invented, of course, but that's not the point. These things were undoubtedly commonly said and felt at the time, and we could easily hear our own version of this list today, involving calculators, word processing, cell phones, social media, etc.

Things are *always* changing, as we know. The point is, we experience that change as an *evolution*, not a *revolution*, so it's hard to keep perspective. It's why we laugh at the list above, but worry about calculators.

And there's the rub for our education systems. In a few years, we will be laughing at today's list, as we now laugh at the ones above. The question is: What will be on the "changing" list *then*, and are we preparing our students for those things *now*?

WHERE DO WE GO FROM HERE?

We have previously pondered (see *Education in an Information Age*, Section 3, e.g.) how our almost instantaneous access to information may drastically change what we can, or even should, be doing in our classrooms. I recently got this same realization/insight delivered from an unexpected source. I found myself in a brief, serendipitous e-mail conversation with Dr. David Sallee, recently retired as President of William Jewell College in Liberty, MO. He related the following story:

"At Jewell, in 2011 I think, we had a faculty workshop to help faculty learn how to best use technology in teaching and learning. In that workshop, the speaker asked us (faculty and administrators) if we knew who Sugata Mitra was. No one knew, so the speaker told us to find out. Literally every person in the room grabbed their phone or their tablet and looked it up. In 30 seconds, someone had the answer.

"Then the speaker said, 'I am really irritated with you people. I am the expert; you brought me here to teach you; I am standing right here; I asked the question, so I probably know the answer, but not one of you asked me.'

"He went on to say this, 'The issue is that they [the students] don't need you for what they used to need you for. Your students don't need you for content. They have all the content they will ever need on their phones. You had better figure out what they do need you for.'

"That was an 'aha' moment for me.... It set us on a different path related to teaching and learning."

As I think about this anecdote of Dr. Sallee's, I'm struck by several quick thoughts. They probably all need further examination:

1. While the anecdote above takes place with college faculty, it's clear that its implications extend to K-12 classrooms as well. Fourth graders can google information as quickly as college students.
2. How do we (continue to?) prepare future teachers for

teaching in this kind of environment, when technology allows instant information? This is both subtle and complicated.
3. As an aside, I'd probably quibble with the leader's use of the word "content" above. Not all "content" in an ideal course is "information." Perhaps this is semantics, but we could probably spend an entire column there.

Finally, I'm struck by the phrase "They don't need you for what they used to need you for." It certainly provides grist for the mill as we re-consider the classrooms of the future *and* the present.

All this should lead us to a crucial question. And it should also prompt an interesting perspective check. The question(s) is/are probably obvious: What *do/will* our students need teachers for? How does this change what happens in the classroom? And, again, how do we prepare for it?

The related perspective check is a little subtler, perhaps. In one sense, thinking about a classroom where technology can provide instant information changes everything about what happens in these classrooms. And yet, in another very real sense, it changes *nothing*. It takes us back to our ever-present question of "what is education?" The goal of education from the days of Jefferson and before was never (solely) to provide tons of information, was it? The broader goal was, and is, to prepare students for life. This is to say, to help produce good citizens who are also articulate, able to think broadly, reason critically, communicate well and fairly, and so on. Without having the burden of having students learn, memorize and/or be tested on readily-available information, we can accomplish those goals even better.

SECTION 9
Politics & Political Issues

INTRODUCTION

—

Other than occasionally grousing about politicians' tendency to try to "solve" educational problems with oversimplified solutions, I try not to skate too close to the political arena in my columns. It's not fun for me, and there's evidence the readers don't like it as well either. (See *Tenure... or Gumballs*, the last column in the section.) Since they are *educational* columns, of course, steering clear is usually not too hard to do.

Unfortunately, there are times that general educational topics *become* political, for whatever reasons. And, of course, there are times when education is directly affected by decisions made by politicians, at both the state and federal levels. In those cases, I never know if it's pretentious to comment or cowardly not to.

Every so often then, I sometimes take a deep breath and dive in. These columns represent times I did just that, for a variety of reasons.

COMMON CORE: GOOD, BAD, NEITHER, OR BOTH?

The story is told that two villagers came before a Sufi judge with their disagreement. The first villager stated his case, and, when he finished, the judge exclaimed, "You're right!"

The second villager protested and carefully presented his side of the case, and the judge again exclaimed, "You're right!"

The first plaintiff, surprised and frustrated, objected and said, "We can't *both* be right!"

The Sufi judge nodded, smiled, and responded, "You're right!"

This is how I've found myself feeling while watching the nationwide battles over "Common Core" and the curriculum, implementation, and testing issues that go with it.

I've deliberately avoided discussing Common Core for precisely this reason. The issues at hand—and there are many—are so complex, that trying to "take a stand" is not only difficult, it's dangerously oversimplifying.

Like the Sufi judge, I have found myself agreeing with points of legitimate value on either "side" of the argument. But I also find myself shuddering uncontrollably from some of the reasoning on either side, for there are serious flaws everywhere, as well.

When one carefully bores through the rhetoric and political grandstanding of all parties, one can find real signs of hope in the important points of each side. On the one hand, one can see a sincere, legitimate, and even exciting vision of the classrooms—and the learning—of the future. Such a vision may seem "progressive," but it is attainable, and it is necessary.

Our students are—always have been—capable of visionary kinds of specific and general learning and growth that prepare them for whatever future exists in their adulthoods of this 21st century. And striving for "consistency" in these issues is simply not the same as the feared "government control" issues that always pop up in knee-jerk fashion.

On the other hand, some of the issues of implementation are troublesome, at best, as is the apparent ongoing fallacy that learning can be measured completely, or even accurately, by "testing." There are assuredly things that need to be addressed here, or they will severely undercut any hope of progress.

These statements are way too general and over-simplified, of course, and more details and specifics should be addressed.

The point for now, then, is this: ***We must find some "common ground" in "common core" discussions.*** Unfortunately for all of us, this has become a political football, in which the arguments on both sides have come to be more about one's politics than what's good for education. Facts are hard to come by, various opinions are biased on either side, there is no reasonable "discussion" and it is rapidly becoming a situation that has NO winning outcome.

Which brings us back to the Sufi judge, and his final insight: "You're right. You can't *both* be right." We must find and take the wisdom from each side of this mess of a discourse and proceed toward the best outcome for our students. Indeed, I can guarantee this: If we don't find some common ground, it doesn't matter which side "wins." If either side *wins* this debate, then we *all* lose, and the entire country—and its educational system—will suffer.

HOME SCHOOLS AND EDUCATION

What an interesting (and increasingly political) phenomenon the "home school" movement has become. It seems that more and more, we hear of parents that are going this route. And not just in individual home schools, but in "home school associations" as well.

I've never known exactly how I felt about home schools in general. My views there are evolving endlessly, but I confess I still don't know how I feel. There are *so many* factors to digest, and so many pros and cons involved here, many of them with huge implications.

On the one hand, home schools really *can* provide a chance to do some valuable things that are no longer possible in public schools. There is much more one-on-one work available (I assume), lessons can certainly be adapted to fit the learners and the situations, and valuable field trips can even be taken. Time schedules can be better made to match the situations and temperaments of the learners, and usually there are not parent/teacher/student problems. In the right situations, home-schooling can be extremely beneficial—sometimes life-saving—for some students.

On the other hand, there are all the obvious (and possibly dangerous) drawbacks. So much, it would seem, depends on the parents/adults who are guiding the learning. Teaching is not an easy task (despite the apparent legions, especially in Washington, who mistakenly think otherwise), and it is surely made much harder by the incumbent demands of home-schooling for knowledge in *all* areas of the curriculum. And this need gets amplified with each passing year in school. Plus, there is also a certain "meta knowledge" that good teachers have about teaching in general and working with students that parents may or may not have.

And even with the educational drawbacks, there are always the social sacrifices inherent in not learning (sometimes the hard way) about interaction with peers. These things are important, it seems to me, but we won't follow that detour today.

And these pros and cons above *just begin* to scratch the surface.

I find it fascinating to look back and see how this movement has caused education to come full circle over the past century or more. In frontier America, what little "formal" education that existed, was naturally done at home. Then, as families (and/or small towns) grew in proximity, they began to hire a "teacher" to work with *all* the children of the area. The *Little House on the Prairie* model, if you will. And things "progressed" from there.

Until now, when we have returned to some parents choosing to leave public schools to go back into the home with their children, for any of a wide variety of motivations, some of which may or may not be suspect or valuable in the long run (another hidden danger).

Perhaps the main take-away for these first thoughts (two more essays follow) is this: At the heart of these discussions we once again find ourselves encountering highly different and evolving views of "education" to many constituents in our changing times. One doesn't leave an existing free public school set-up (do they?) unless one has begun to view the broad definition and even purpose of "education" differently, whether it's related to curriculum, values, or something else.

The next selection follows up on these thoughts with a rather radical story, and then we try to begin to tie it all together.

HOME SCHOOLS 2: GRIST FOR THE MILL

Previously, I carefully waded into the "home school" arena, not without trepidation. But, in truth, I did so as a launching pad for this set of thoughts.

I'd like to share this story, as it happens to intrigue me in varied and nebulous ways. But I'd like to tell this story without any of us, *myself included*, jumping to *any* general conclusions of *any* kind. Okay? Pinkie-swear? I'll agree if you will.

We'll start with two quotes, the first by Mark Twain, the second by a former colleague, from the last line of a recent op-ed piece:

1. I have never let my schooling interfere with my education.
2. In the absence of "formal" schooling, create lessons from life.

Both quotes relate to our story, but possibly in a way neither author quite intended.

Our son Adam recently attended a wedding back in the Shenandoah Valley of Virginia, where he used to live and work. He was housed for at least part of his stay by one of his favorite set of friends, Jason and Janelle, who have two incredibly bright daughters. Kali, the older at twelve, has a love of all things mathematical, which she shares with Adam. (Who knows where *he* got it?). Adam also has (inherited?) a passionate-but-unique love for education, learning, and growth, though this is not his current vocation. His perspectives are often interesting. (Witness, for example, his comment in the book's Prologue, which was a seed for much of the thinking that lead to these columns, and therefore this book.)

During the visit, the two of them (Adam and Kali) were discussing a diverse collection of math problems from old middle school contests, which he had previously sent to her, in anticipation of his visit, and to fuel her math passions. He was most impressed by her work on some of them, despite her not having seen some of the material formally. She didn't just "get" all the problems, she knocked them out of the park, some with quite unique/clever approaches. (Compare with ***Section 7B: Mathematical Thinking.***) Adam was delightfully amazed, especially since her motivation was all internal.

He was relating all of this to us on the phone after returning from the wedding. My wife was captured by *his* amazement. Suspecting something further, she asked, "Do they home-school the girls?"

There was a significant pause, a deep breath, and then his reply: "Well, no, they try to *un-school* them."

I sensed his meaning, but he continued with an example. "One of the days, we were out in their garden and Jason was showing me some of their current work. [Both Adam and the friends work in the sustainability/permaculture movements.] Kali was nearby working on her own area of the garden, in which she has *already* created a new, improved, hybrid strain of corn for that area! She had some questions and called for help. We went over, and Jason answered her questions and made suggestions."

"As we returned," Adam continued, "Jason winked at me and said slyly, 'and to think she could be at school *learning* right now.'"

I'll add here, FYI, that both daughters are given choices: if they ever want to go to public school at any time, they may. (The younger, at seven , is leaning that way.)

Is this story connected to education in general? Absolutely. Do I want to elaborate today? Not really. And, remember our agreement: I certainly don't want to extrapolate from one intriguing story. I'm not *advocating* anything here.

But I'm reminded, as our opening quotes suggest, that in the real world, *authentic learning* A) doesn't always come *in school*, B) even then, can't always be scheduled to fit our convenient agendas, and C) often happens when there is a need that dictates it.

HOMESCHOOLING 3: WRAPPING UP

—

After the last two columns related to homeschooling, and the various responses I received, I have re-affirmed the age-old, slightly paraphrased maxim: "the best way to expand your thinking about something is to write a column about it."

I had an interesting variety of responses and experiences arise from the columns. Among the highlights:

1. Communications with an educator connected with an online homeschool curriculum provider that also provides classes for individuals or associations. (I would call it an online private school of sorts, but my correspondent would disagree.) We had some *great* discussions and found ourselves in basic agreement on most points related to educating students in general.
2. A request to be a guest on a local talk show to discuss homeschooling. This one scared me mightily, as I certainly do/did not claim to be an expert on homeschooling. But I survived the experience, I think, and the discussion helped me to further clarify some individual thoughts as well.
3. A very *articulate* set of thoughts from friend and former colleague in mathematics education—Dr. Terry Goodman*—who, having now retired, has partnered with his wife, a retired public-school teacher to home-school their grandson. (* We met Dr. Goodman in **Section 5, *Covering Material*.**)

I'd like to elaborate on #3. Terry's responses were most enlightening as they combined his career experiences in higher education with his current experiences in homeschooling. I've had to edit much of his response in this space, but I think I've preserved the core. He begins: "Having worked with public school teachers (pre- and in-service) for 40 years and strongly believing in the value and importance of public schools, it was with mixed feelings that Teresa (a public-school teacher) and I decided to homeschool our grandson."

He then highlights an important advantage in their experience, namely that "we can more easily and consistently identify and meet [our grandson's] needs."

But his strong caveat: "Now that we are homeschooling our grandson, we are even more convinced that being the homeschool 'teacher' is a very challenging task. Choosing curriculum materials, especially in areas that are not our 'specialty' areas requires careful thought and research. Building in appropriate and consistent extracurricular/social activities requires even more time [as we] construct a schedule and curriculum that takes into account our grandson's learning disabilities."

All this brings me back to my post-column experiences. These interactions allowed at least two things to happen for me: I was reminded what a difficult, though timely, topic this is to discuss. And, I was helped in clarifying at least *some* of my thinking in this area.

That increased clarity, or insight, *for me* is this: The decision to homeschool is (or should be) *all about the student(s) in question,* and not so much about parental beliefs or fears. (So much unsaid here.) Choosing to homeschool a child *can* be a valuable—and viable—option for parents (or grandparents) if it helps the student who may have special need(s) and/or needs help in some area(s). At the same time, it is not a decision to be made lightly and comes at a *very high* cost for the [grand]parents in terms of time, finances, organization, and lifestyle. Parents who make this choice should be prepared for these demands. Homeschooling is *not* an easy option, especially if it is to be productive and beneficial to the student(s) involved.

AN EDUCATIONALLY TROUBLED NOMINATION

Background: After the 2016 General Election, one of the most controversial Cabinet nominations was that of Betsy DeVos for Secretary of Education. This is the column I wrote during that process.

As a matter of perspective, I used to joke that "everyone considers themselves an expert on public education because each person was a student." Now that joke has a cruel and frightening twist. We apparently now have someone who is considered an expert for the opposite reason.

Short detour: Earlier, we discussed home-schooling topics for some time. I believe our discussions could be summarized thusly: Home-schooling is neither bad nor good, in and of itself. Home schooling and public schooling can each do some things the other cannot, but the choice for either should be well-informed and not lightly undertaken.

I hope that's fair, because I'd like to take the same balanced position with charter schools, without having to take three columns to do it. Specifically: Charter Schools are neither bad nor good by themselves. Charter schools and public schools each have strengths and weaknesses. (Note: Whether charter schools should use public funding—through vouchers or other means—is an important-but-separate discussion even more related to politics than education.)

To be automatically opposed to *either* charter schools *or* public schools (or home schools) without doing one's homework is risky at best. Yet we are perilously close to having that happen on a national level.

Consider for the moment a person who has, apparently, A) *never* been a student in public education, or put any of her children in public schools, and B) *never* been a teacher or school board member or had *any* contact with the workings of public schools.

Such a person exists and has now been nominated for Secretary of Education.

To add to this mix, this person, and her husband, have also C) spent large sums of their personal wealth fighting public schools in general, and D) simultaneously and mysteriously (with more of their money) fighting *all* efforts at regulation and accountability for those charter schools they have helped create. They seem to oppose public schools on principle, while opposing accountability for charter schools.

Regardless of your political leanings, and *regardless* of your opinions of and preferences for public, home, and/or charter schools (or even the vouchers question), are these the kind of narrow views (and actions) we want in the person who leads the Department of Education?

Do not take my word for these facts. I urge you to "fact check" and do your own homework. Form your own opinions. (For my source[s] on this column, feel free to contact me.)

At the risk of being redundant, I am neither a fan nor foe of charter schools, *per se*. I see possible strengths in the concept of charter schools, as well as the ideas of home schooling, for those who choose them. But that's not the point. The Secretary of Education must be able to facilitate and promote the best levels of education for *all* our nation's children, regardless of the vehicle. After all, *education for all* is a cornerstone of our democracy, set firmly in place by our founding fathers, especially Thomas Jefferson.

This is not really about politics. What would exacerbate this situation even more is to let it get (or stay?) mired in politics, as opposed to education. Democrats should not oppose this nomination for the sake of politics alone. Republicans should not favor this nomination for the sake of politics alone. Everyone who is interested in education should oppose this nomination *for the sake of education alone.*

WHAT IS THE GOVERNOR DOING? AND WHY?

—

Background: In late 2017, much controversy erupted (and continues, as of this writing) in Missouri, when the new Governor attempted to fire the current Commissioner of Education (a position he has no authority to hire/fire directly), by continually appointing members to the State Board (without confirmation), and then withdrawing their nominations if they did not seem willing to follow his desires in the matter. The next two columns were written at that time.

I may disagree with Governor Greitens' apparent ideas on education, but I will certainly concede and defend his right to have those views, and even to try to convince others of them.

At the same time, there are (at least) two things that are deeply troubling about what is going on in Jefferson City [Missouri's capital] with respect to his attempts to replace the current Commissioner of Education.

First, the Governor seems to be unwilling to share, let along discuss, his ideas with current educational policy-makers to work together to reach the common goal of what's best for Missouri's children.

Second, and much more troubling, he is blatantly working to circumvent checks and balances that are in place to separate educational decisions from political whims. Such separations are in place *precisely* to avoid these politically motivated power plays. Yet he proceeds, apparently uncaring about either the appearances of impropriety or the results of such machinations.

I am aware that I may be guilty of being an educator running amok in an unfamiliar field of politics. But I do so with a grin of irony, fully aware that such a practice exactly mirrors the Governor's actions, in reverse.

Let's pause, then, and look at the educational elephant in the room. All the rumors suggest that the Governor's not-so-hidden goal in all this is to bring in an out-of-state Commissioner who is an avowed advocate of charter schools.

I will not bash charter schools, per se. I have stated before that good charter schools can conceivably achieve some things that public schools cannot. (The reverse is also true, of course.) Instead, let's look at just one cautionary snapshot, among many.

In the early 2000s, the state of Michigan opted for a complete "choice" model, with charter schools and no district lines. This change was due, in large part, to the prodigious efforts (and enormous financial backing) of Betsy DeVos, current Secretary of Education, and her husband Richard.

In the decade that followed, Michigan's standing on national tests dropped from the middle of the pack to near the bottom. In 2003, Michigan ranked 28th among the states in fourth-grade reading and 27th in fourth-grade math (as measured by the National Assessment of Educational Progress). In the 2015 results, Michigan had dropped to 41st in reading and 42nd in math.

To be sure, this is just one example, but it is noteworthy and pertinent. It should serve as an indicator of caution for two reasons. It shows that charters frequently do not live up to whatever potential they may have. And it also warns us to beware of hell-bent efforts to put them in place without careful examination.

Let's learn from Michigan. Let's not rush into an action that may not be good for Missouri *or* its students. And let's certainly not rush into it with brute force. Discussion, of course. Debate, sure! Both are called for. And both are far better than unilateral attack, based on individual prejudices, in direct defiance of existing State structures.

We should not ask Governor Greitens to abandon his opinions. But we should all call on him to slow down and abandon his current questionable and disappointing approach. (Can he *convince* others of his ideas, rather than bulldoze them?) And, we should ask the governor to respect and work within existing structures. We expect our school children (and our citizens) to do this; surely, as Governor, he can lead the way.

THE BROWN SHOES AWARDS

"Did you ever feel like the world was a tuxedo and you were a pair of brown shoes?"

Folks my age may remember comedian George Gobel. Once, as a guest on the Tonight Show with Johnny Carson, Gobel (appearing on the same show as Bob Hope and Dean Martin) slipped this line into the conversation, cracking Carson up in a way that was unusual.

I remembered this image recently, while thinking about the current State School Board controversies in Jefferson City. I decided it would be fun to start "The Brown Shoes Educational Award," to be presented to those whose *actions* seem to clash with accepted sensibilities, at least in education.

I stress that this B.S. Award is presented primarily for actions, and not necessarily personalities. Also, in the interest of full disclosure, I note that the Awards Committee consists of one person only, and his views are certainly subjective.

Perhaps you've deduced the Winner of this First Round of awards: Governor Eric Greitens. The voting wasn't even close. His willingness to blatantly, maliciously, and perhaps illegally sabotage existing state policies and procedures to further his own unspecified educational aims is not only shameful, but borderline arrogant. His actions could win both the Educational and Political B.S. Awards. Hopefully the legislature will present him the latter.

Particularly galling is the governor's unwillingness to share or promote his views on education for discussion. We have heard he thinks teachers are underpaid, with which most of us can wholeheartedly agree. We have heard that he thinks administrators are overpaid, which is much more debatable—that is, if the governor allowed debate. His apparent willingness to believe that his views alone are worth the travesties he has perpetrated are more than enough to color his shoes brown.

The Runner-Up Award, which comes with truckloads of

disclaimers, is shared by the five (of the ten appointed) commissioners who were willing to bow to the governor's wishes, without further examining the issues and apparently without caring about the questionable means being used to attain the ends.

PLEASE: Most pertinent of the disclaimers is this one: I have no problem whatsoever with our own [local appointee]'s reasons for accepting her appointment. (See ***"Why I Accepted...,"*** *News-Leader, Dec 3, 2017, page I1.*) I agree that dyslexia is real and can hinder learning. Apparently, (this local appointee) agonizingly decided that her chance to advocate for this issue was worth the extremely high price she paid to do so. And while I may disagree (and worry that she may have hurt her cause), I will respect that particular part of her decision.

Nonetheless, it is still the case that all five new members (none of whom are yet confirmed), deliberately chose to sacrifice open discussions, existing structures, possible legality issues, and certainly common sense, to advance the governor's cause, while condoning his methods. For that, their shoes are distinctly brown.

When it comes to tuxedos, black shoes are rarely noticed. But, in this case, there are also Black Shoe Awards, and there are Dual Winners. Sharing one award are the *other five* Greitens appointees to the Board, who refused to automatically vote with the governor's wishes without further examination of the issues and circumstances. (As a result, their brief tenure on the Board is over.) The other Award goes to Sen Gary Romine (R), who has acknowledged that "this is just not proper procedure," and will work to reestablish the Senate's state-mandated role in these appointments through the Senate confirmation process.

If the shoe fits....

THINKING ABOUT TENURE

Background: This column (and the next) was written during a time when a bill proposing the elimination of tenure at institutions had been introduced into the Missouri legislature.

Retired executive Jack Welch once said "Tenure is a terrible idea. It keeps them around forever and they don't have to work hard." (One assumes he is speaking of professors.)

It's dangerously easy to buy into this reasoning. But if one is not careful, it's also easy to believe that public teachers work *only* from 8 to 3 on weekdays for nine months a year. (Or perhaps that ministers only work on Sundays, and "don't have to work hard.")

Tenure in higher education can easily be misconstrued. Therefore, it has gradually become more controversial. Politicians in Wisconsin have managed to gut the idea, and bills are pending in both Iowa and Missouri to eliminate it.

Let's calmly explore some facts about tenure. For today, at least, I'll restrict my remarks to tenure in higher education.In the first place, tenure is not automatic upon hiring. At most, if not all, institutions, it takes *seven years* for tenure to be granted, and then not easily. Do detractors know this?

This gives the institution a great deal of time to evaluate a professor and his/her work. Seven years is a sizable chunk of anyone's early career, and stereotypically, new professors work *much* harder than they should during those years to try to "over-insure" that the granting of tenure later will be an "easier" decision.

So, why is there this assumption that a professor will "sit back and do nothing" after tenure is granted? Most professors love the research, teaching, working with students, and community service in which they are involved. And while they may (probably should) "let up" a little to return to a saner life, they rarely sit still. Why would they? Their profession is not "just a job" to them. (Besides, raises are not automatic.)

Like it or not, higher education is a profession where the ability to be protected for *doing one's job* can easily be needed. Here's a case in point: Representative Rick Brattin from Cass County, who has introduced Missouri's anti-tenure bill has said "If you're doing the right thing as a professor and teaching students to the best of your ability, why do you need tenure?"

Think about that statement. For starters, he appears to have ignored the important research component of many professors' jobs. Second, please tell me: *who* gets to decide what is "the right thing?" What about a professor that is teaching, say, an established fact of science that Mr. Brattin may not like, in a legislature controlled by his party?

This is *precisely* why tenure was born, and is still needed. Mr. Brattin's statement is a excellent argument against his own bill.

In higher education, especially at bigger universities, part of the goal is to help find *new* knowledge, as well as disseminate existing knowledge. This *can* be controversial, and has been in the past. Would Mr. Brattin, living in Galileo's time, have wondered if Galileo's then-highly-controversial teaching about a sun-centered universe was the "right" thing to do? Does Mr. Brattin now consider science-based teaching on climate change, the "right" thing, *regardless* of his own (or his constituents') views?

It is important to remember that one of the goals of education—the ongoing search for truth and knowledge—is not just an idealistic whimsy. It is important to us all. It must be allowed to continue (and to be debated and tested) in an environment that is free from the pressures of politics, religion, and/or society. If this ability is restricted, truth suffers. Tenure exists not just to protect the professor, it exists first to protect the truth.

TENURE... OR GUMBALLS?

After my last column, I was gently and good-naturedly chastised by a "faithful reader." The input was interesting.

Essentially, I was chided for my two recent columns which dealt with 1) the Education Secretary nominee and 2) the topic of tenure. The complaint was not so much for content (there were no signs of agreement or disagreement), but rather for the *nature* of those columns.

I'm mostly paraphrasing, but this reader preferred my normally lighter columns (which are apparently "a fun distraction-like a crossword puzzle") to the heavier content of the two in question. Specifically mentioned as preferred was a real-world gumball-packing problem discussed last year. (See Section 7 and the last two selections.) By the way, I liked that one, too.

I appreciated the comments (and especially the calmer, non-adversarial nature of the note). Truth be told, I'm in partial agreement. Personally, I dislike *writing* the columns that skate close to the thinner ice of the political arena (this section notwithstanding). Normally, I can avoid that.

But much is happening these days in the world of politics (now *there's* an understatement!) and some of it can greatly affect the world of education. Like all matters of education, these things can greatly affect all of us. So, some of these ramifications need to at least be observed and/or commented upon as they occur.

All that said, let me use my remaining words to pitch out two more important points on the tenure theme, before we skate away from it entirely. (Sneaky, eh?)

First there's this idea that "in the academic world, you can get away with literally anything," a quote directly from Rick Brattin, the legislator from Cass County who we met last time. Such a thing is simply not true, of course. There are procedures in place for most eventualities, and safeguards for all concerned. Yes, it is hard to "fire" a tenured professor. It should be, for all the reasons discussed last time.

Nevertheless, let's interject some perspective here. All things considered, I don't think I'm out of line to suspect that it is probably no harder to be fired for a heinous act of some kind in the academic world (tenured or otherwise) than it is to be removed for the same act in Mr. Brattin's political arena. I'm *not* being sarcastic or derogatory, just realistic. Cases in point abound, especially nationally, and especially recently. Pick your own. I'm just not sure Mr. Brattin should be so dismissive.

Finally, there's the not-quite-so-apparent economic dimension to Missouri's HB 266 (the tenure bill in question). Suppose you are a promising young researcher, or a gifted teacher of students, or both, just finishing graduate school or looking to make a move of some kind. Suppose you are offered two equal university positions, one in Missouri, with HB266 in effect, and one in another state that has *not* eliminated some job security in this world of changing political influences. Which will you take?

I suspect the answer to that question is obvious. How, then, will that answer affect Missouri, in both the short and the long term, if it can no longer attract the same quality of professors it once did?

So, I return to my slightly annoyed reader from last time. Hopefully, we can *all* return to the columns which are more fun to read, as well as to write (bring on the gumballs). But I have this uneasy feeling there may be more detours ahead. If so, keep the comments coming.

SECTION 10

A Potpourri of Flavors

INTRODUCTION

We shouldn't expect everything to be able to be categorized, should we? In some respects, I'm lucky to have been able to find nine broad (albeit overlapping) themes into which to roughly fit the selections in this book.

In this concluding section, then, we visit about a small olio of different topics. A couple of these (the first two) are indeed related, but the others represent topics or themes about which I have only written once so far. These range from parents to my wife's view of my profession, with some others in between. A couple of them *hint* at controversial topics, but since I don't try to dictate a point of view, I suspect you'll find them palatable.

So, sit back, pick the ones that interest you, and dive in. I hope you'll enjoy the assorted flavors.

THE THREE A'S–PART 2

In an earlier column, I shared some very quick scenes from my life's movie. These scenes all related to the role of athletics in my growth, in more ways than one. One of the scenes involved meeting a colleague at a conference who suggested that we should pay more attention to "The Three A's" in school/life: Academics, Arts, and Athletics. I'd like to play with that notion for a while.

One of the things I like about this suggestion is the obvious analogy to the "mind/body/spirit" trilogy that seemingly all health experts, psychologists, and spiritual leaders seem to suggest is important to strengthen and keep in balance in our lives.

Does that analogy fit in our education system? Does it provide a broader view of "education" than we're used to? After all, the "three R's" most of us grew up with (reading, 'riting, and 'rithmetic) mostly focus on Academics, don't they?

I don't think this is a new notion. Indeed, if anything, I think we used to have that more balanced view of education, and we're getting *away from* it, rather than moving toward it. Clearly, we still focus on the "head": academics. But for whatever reasons (all too often financial ones), schools are de-emphasizing "the arts" (the spirit), and "physical education" (the body). That may be penny-wise, but it seems dollar-foolish.

[By the way, let's clearly distinguish here between physical education in a broad sense, including sports, and its burly cousin "big time athletics," which is usually more about money than sports and is often a case of the cart pulling the horse.]

This increasing de-emphasis of the body/spirit part of our students' education is alarming, and seems to be detrimental to all concerned. We've all read the studies of the obesity problems that are so prevalent with our youngsters. Would increased awareness of the importance of healthy fitness (to students *and* parents) not be a wise topic to reinforce in our schools? Perhaps this also means an overhaul of "required PE classes," but, as a society, we need to urge students to more *activity* and better physical health.

Likewise, the arts. Not every student will become an artist or a musician or a poet or a writer or a fill-in-the-blank. But some will, and more *could.* And they *must* have that outlet and opportunity. We've said this before, but other cultures and societies, past and present, value their poets, their writers, their artists, their musicians above all others for their ability to touch us, to awaken us, to keep us attuned to what is real and what matters in life. In other words, to nurture our *spirit.* Can it be coincidence that the decline of "the artist" (and his/her value) in our society these days seems to correspond to the decreasing emphasis in these things in our schools?

Academics. Athletics (in the purest sense). The Arts. All three are important. All three affect us greatly, as well as our ability to "make a life" as opposed to only 'making a living.' To forget that fact, or to get the three out of balance, is dangerous—in our lives, certainly, and *especially* in our schools. Whatever we call it, perhaps we should continue to remember that the idea of Three A's continues to be as, if not more, important as/than the notion of the Three R's.

PLUTO, SCIENTISTS... AND THE ARTS?

As I write this, the New Horizons spacecraft has just finished its "fly-by" of Pluto, capping a 9.5-year journey that has delighted the NASA scientists, and also captured the imaginations and the hearts of most of the rest of us.

Another incredible achievement of science & space exploration. That such a feat could be accomplished, when it takes nine hours to even exchange information with the spacecraft, is amazing. So, here's to the scientists, and here's another commercial for the importance of math and science in our educational curriculum.

But perhaps you expected that from this corner. What else would I say, right? Perhaps you won't expect what follows.

The day before the fly-by, columnist Charles Krauthammer spoke to this topic in his column. I don't often agree with Krauthammer politically, but this column was most interesting. He spoke to *why* we would undertake this project, and gave two reasons:

"First, for the science, the coming avalanche of new knowledge." As we know, Pluto is the remaining mystery of the original nine planets (it was still a planet when the probe was launched). We will be gathering information and knowledge from this probe for months, even years.

"Then there's the romance." I didn't see this one coming from this columnist. He continues, eloquently speaking of the pull of such explorations, and essentially, the mystery of the universe in which we live.

Space exploration has always seemed to be the one area where the science of the universe seems to blend with the mystery of the universe for all of us. So, let's make an educational connection, as well.

We have heard it so much: "The goal of education is not just to prepare us to make a living, but to enable us to make a LIFE." We usually say this referring to a healthy balance of economic and quality-of-life issues.

But I'd like to recast the statement in another light. I think it is also about the values of science *and* "the arts." Unfortunately, we continue to live in an era in which the former is highly valued, and the latter is under-valued and often cut from school curricula, due to budgetary considerations. This imbalance is not healthy.

No matter how much science tells us about how the universe works, there will *always* be more that we don't understand, and at which we continue to marvel. Regardless of our religious beliefs—or even, for that matter, our lack of them—there *always* remains a grand, unfathomable mystery to the universe in which we live. We should cherish that awe. How we are affected by, react to, and interact with that Marvelous Mystery affects the quality of our life. ***It has* always *been the job of the poet, the artist, the musician to remind us of that mystery and to put us in touch with it.***

Without the scientists, we would lose much of our ability to explain the universe in which we live. But, without the artists we cut off a significant portion of our preparation for actually *living* in that universe. We must, it seems, continue to nurture both the sciences and the arts in our schools, for they both prepare us, in equally important ways, to function in our everyday lives.

Here's a toast to the scientists, yes. But let's also toast the artists, as well. And let's continue to develop both the sciences *and* the arts in education. To ignore either is to invite peril.

PARENTS & SCHOOLS–A DELICATE BALANCE

It's hard to go very long these days without seeing a story, locally or nationally—about parents and schools being at odds over one or more issues. And often teachers are caught in the middle, *especially* if they're both teachers and parents!

Like other parent-educators, I've been on both sides of these debates. I've been the strongly-concerned (and outspoken?) parent who thought the schools were missing the forest for the trees on an issue that concerned one or more of my children. And I've been the educator who's been in favor of things that often get parental resistance (appropriate usage of calculators in the elementary classroom, for example) and wondered why the parents couldn't see the big picture as the education process struggles to keep up with changing times. (And ask me sometime about the FBI agent who disputed his daughter's grade on a final project.)

Obviously, this parents-versus-schools issue is another arena with no easy answers (which is always part of my point, of course) and obviously, it's an arena that requires good honest communication with and among all parties. Perhaps a good starting point would be with a perspective check that's probably just as obvious, but is one I haven't seen mentioned a lot in all the letters, commentaries and media reports.

Perhaps it would help to continually remind ourselves that regardless of all else, parents, educators and administrators are in fact *always on the same side*. They/we all want, in the long run, what's best for our children, our students, our learners, our future adults. They/we always want our children to become good thinkers, good problem solvers, good communicators, and good citizens. Clearly, then, the disagreements come over the issue of *how* to secure that goal. And, in that light, the disagreements can get less intense.

Somehow, at least for me, it seems easier to get to that "honest discussion" stage with someone if we're both remembering (and acknowledging) that we basically want the same thing. That doesn't make the original concern go away, of course, but at least it changes the focus: we're starting from a place of agreement and moving from there to the (often difficult) question of how to reach that common goal.

Another thought struck me as I wrote about having been on both sides of these situations. As I tried to look at myself objectively, I couldn't help but notice the issue of ownership creeping into the picture. I wonder: Would the issue at hand be quite so crucial if it weren't *my* child that was being affected? Would it be easier to see the truth on both sides if it weren't *my* idea/position/belief that was being questioned by a parent?

It's certainly fine to be invested in one's ideas or opinions, but it's amazing how far a little detachment can go (in most situations) toward providing some perspective and getting both sides to a win/win quicker. (Would that politicians could remember this, eh?)

As long as there are parents and schools, there will be differences of opinion about educational issues. This is probably as it should be. It shows caring for our students on all sides and from a variety of perspectives. But when those disagreements arise, does it have to be parents *versus* schools? Why can't it be parents and schools working as a team toward their common goal?

MY WIFE, MY MATH, AND... EDUCATION?

No, that's not the start of a classic bar joke, but there *is* some humor involved. And we *will* get to some educational observations.

Throughout our nearly-half-century of marriage, my wife Pat has dealt with my mathematics education career with a wide variety of opinions, and even boredom. (Attend a math conference with me? She'd rather have a root canal.) It has led to some funny incidents and standing jokes. Perhaps the funniest (now—not then) was when she accidentally threw away the equivalent of an early part of my dissertation!

Understand that Pat is highly intelligent, full of personality, and, before she retired, had ascended to amazing heights in the business world. But her success has led to some of our interesting situations.

Here's one of them: For years, when the women with whom she worked and mentored could not quickly figure 50% of something, the standing joke is that she would often get mad at *me personally* for not teaching students better.

That's not logical—we both know that—but, on one level I can feel her pain. And, if we look at my profession, I can *almost* see her point. *Why* should any high school graduate *not* be able to instantly at least *approximate* 50% of something (or 15-20% for a tip)? And isn't *someone* to blame for that?

Well, no, but the deeper question persists: Why are some universally-taught math/arithmetic skills so universally forgotten not long after school days? This phenomenon exists in other subjects too, of course (when was the Battle of Bull Run, or what's the boiling point of water?), but it seems to be wider and more pronounced in math. Why?

If you've read more than one of my columns, you know this is not the place where I reveal *The Answer* to that deep mystery. I'm not sure there is (just) one, and exploring further could take a dozen columns.

If, however, you want my "favorites" of the "contributing factors,"

I'm willing to mention at least one. This is *not* universal, of course, but it seems to be *too* common, which is "our" fault.

Many important educational skills, especially mathematical ones, aren't taught as *useful tools*, along with why they should be in a person's Tool Bag for Life. (How does "moving a decimal point two places" to the right—or is it left?—in/of itself sell the importance of percentages?)

Too often, these handy tools are presented A) as rote routines to be memorized in isolation, and B) in a boring manner (see A).

We have space for one quick, isolated example out of many: Take the Pythagorean Theorem. (I know, please! And take it far away.) For some reason, this seems to be one of the things many of us remember hearing about: Aha! It's $a^2 + b^2 = c^2$. And that's partly right, in a partial and isolated context sort of way.

But there's so much more fascinating information here. It's really *quite an amazing* relationship, with beautiful visual effects and marvelously fascinating extensions. This incredible result should be taught with exclamation points, not sleeping pills.

We've only begun to scratch the surface on this one. *Mea culpa.* We'll be back in this neighborhood soon. In the meantime, if you're near Pat and need to take 50% of something, *please* get it right. My well-being depends on it.

EDUCATION AND THE FOUNDING FATHERS

Recently, I encountered a 2015 article/blog entitled "***7 Things our Founders Believed about Education.***" The author, David Akadjian, builds his list using various quotes—primarily (though not entirely) from Thomas Jefferson, John Adams, and Benjamin Franklin.

A brief diversion. Invoking the opinions of even some of "the founding fathers" should probably be an exercise in caution. It's being done so much these days, and in such a haphazard fashion, that it's hard to know what's biased "spin" and what's productive insight.

On the one hand, hearing and reading actual quotes of some of our founders can be instructive. Voices separated from us by almost 250 years can provide perspective and insight into bigger picture intentions. On the other hand, times *do* change, and it's wise to ask how original intentions fit into our society over two centuries later. (One only need witness the ongoing debate on the 2nd Amendment to see this in action.)

With those perspective reminders, we return to the article. Using several direct quotes taken from correspondence, speeches, and writings, the author makes a case for seven core beliefs about education shared by our founding fathers in our early days as a country: 1) Education is critical for democracy and 2) for avoiding an "aristocracy of wealth." 3) It (education) should be available to all, 4) should be free from religion and ideology, 5) should be equal for all citizens, and 6) should be public. Finally, they believed 7) that the investment is worth the cost.

I guess I would say that I still basically believe the spirit of all seven of those points to be important cornerstones, or at least good starting points for reminder and discussion. But my own belief is not so much the point.

The point is "what do *we* still believe?," both individually, and as a society. Have times changed some of our beliefs? Has it changed some of our actions? Amid the current political upheaval and apparent attacks on some of these principles from positions of authority (including, perhaps ironically, the Secretary of Education), what do *we* still think is important? What do "we the people" believe in enough to be moved to discussion and/or action?

Do we still believe, for example, that a good public education is worth the investment (Item 7) and are we willing to match that believe with action? Do we still believe in avoiding an "aristocracy of wealth" (Item 2) and if so, are we succeeding? Do our beliefs match the current political realities, and if not, what are we willing to do about it?

All seven points raise interesting questions. All seven could, maybe even should, produce disagreements of assorted sizes. All seven might require an agreement on terms to reach productive consensus. (Item #4 alone—free from religion and ideology—could be a powder keg, if not handled delicately and maturely.)

But, perhaps the point is that *all seven* principles—or similar ones—*should* be constantly and calmly discussed. Amid all our constant and conscientious planning for the immediate future, we should know what we feel about these overriding principles and let our broad beliefs guide us into the future. We owe it to our future citizens, and we owe it to our founding fathers.

*Thoughts similar to those in this column may be seen in Section 1 (**What is Education?**), Section 6 (**Administration and Support**), and others.*

OPPORTUNITIES TO THINK?

I attended a handful of institutions of higher learning in my early academic sojourns. As a result, I have received tons of alumni magazines over the years. These come in a variety of formats (many of them now digital), but they always seem to have the same feel and plot-line: Stories about various faculty members, current students and an alum or two, lists of updates of graduates (organized by year), often a message from the President, and of course the omnipresent opportunities to support "your" institution financially.

Occasionally, there'll be a surprise.

Years ago, I got such a surprise, when the Winter 94/95 Alumni Magazine for DePauw University (Greencastle, IN) arrived. I was about to toss it when an opinion piece near the end leapt out at me and grabbed my attention. The article, a book excerpt, was written by an alumnus named Richard Peck, and it spoke to me for a couple of reasons.

The article itself was about modern-day censorship in schools. Looking back, it's as current now as it was then. I don't want to get too sidetracked here, but a quick synopsis is in order.

The piece pulled no punches as it spoke about censorship *both* from the fundamentalist right *and* from the liberal left, with interesting hard-hitting critiques for each. Primarily, however, the excerpt seemed centered on parents who try to dictate or suppress local educational curriculum. A sample quote: "Censorship [in this area] isn't about books; it comes from redirected parental fear... We'd be much nearer nirvana now if parents respected schools more and feared their own children less."

Whew! As I say, he pulled no punches. For my own part, I wasn't sure then how I felt about some of the quotes, indeed the entire article. Certainly, the piece was (destined to be?) controversial, and I suspect—I don't remember—that it drew alumni letters of both praise and criticism in the next issue.

Topic aside, however, that's *exactly* the other reason the piece drew my attention. The fact that a potentially provocative piece would appear in this communication with alumni was a refreshing change for me at that time. Certainly, DePauw was not afraid to make us think. [Probably there was an "opinions of..." disclaimer—again, I don't remember.] Unfortunately, I know from experience, there are colleges who wouldn't come close to allowing anything thought-provoking in their literature, or even more sadly, in their classrooms.

And, when it comes to that, maybe alumni literature is a good barometer of the learning atmosphere on campus. If they're still encouraging thinking in their alums, you can bet it's happening on campus as well. Another quote in DePauw's literature went like this: "Our dedication must always be to ensure that this is a community of learning... a community of adaptors not of adopters: a community that learns not only how to assemble knowledge, but also how to understand it." Certainly, DePauw was showing they believed that.

When all was said and done, I remember loving the way that article made me think. It tossed out ideas from a perspective I hadn't thought of and caused me to examine what I really believed. Because of that, I came to know myself a little bit better. Isn't that part of what education should be about?

NOTES FROM THE RIVER: TRAVEL AND LEARNING

—

Francis Bacon said, "Travel in the younger sort, is a part of education; in the elder, a part of experience." I would rephrase that: Travel, for all ages, is both experience *and* education.

Why am I waxing so philosophical here? It's because of the Big Missouri River Adventure I was on, as this piece was written. Beginning late summer 2016, I began a 51-day trip (by car) along the Missouri River from Three Forks, Montana (the River's official, not actual, source) to north of St. Louis, where it empties into the Mississippi. (See Appendix 1 for more details.)

Columns for dates in that span were sent in ahead of time. I didn't quite realize I'd be writing one "from the road/river" as it were, but I became inspired to talk more about travel, learning, and education. From here, I pick up the original column, already in progress:

I'm learning a lot out on this trip already. Oh, yes, I'm learning some new facts, lots of them. For instance, Lewis and Clark *County* in Montana (where Helena, the Capital, is located) consists of *over* 3300 square *miles*, more than the area of Rhode Island and Connecticut combined. I used to think "Big Sky Country" was a slogan. Now I think it's an understatement.

And I'm obviously learning more about Lewis and Clark than I remember learning in school, and it's fascinating. And it's much more *real* than it was in the classroom.

But I'm learning more than facts, too. It dawns on me that I'm not just exploring the Missouri River. I'm exploring America itself, or at least an important segment of it, and its people. I encounter new people—not online—either in passing or in depth, every day, and 99.37% are great people. Like many of us, I suppose, I've grown cynical over the years of what I view as "false patriotism," as spouted by many of our politicians. But, I tell you: I'm barely half way done with this trip, and I'm feeling better—by far—about America and its people than I've felt in years.

And there's more. I'm also discovering that, like any adventure, I'm exploring myself as well. Just yesterday (as I write this), I got to ride through the country where "Dances with Wolves" was filmed. It's so magnificently *big* and beautiful as to be awe-inspiring. *Continually.* My host's father says, "Folks who say there's nothing to see don't know what they're looking at!" Why is it that these views, and more, touch me in places I'd almost forgotten I had?

Okay, Okay. Not everyone can (or would want) to engage in such a long travel trip/adventure. So, let's bring this back closer to home. Before I left this summer, there were some discussions in (another columnist)'s column—and with his readers—about educational things students could do this summer, while not in the classroom proper. There were some good ones. Summer's almost over, but let me add another, and not just for summer. Take a trip, even as short as a day. Your area is likely FULL of marvelous adventures within a day or weekend away. Find one you and/or the kids like and go there. It's good for the mind, the body, and the soul. It's education!

For more on this trip, and the book the it spawned, see ***Appendix 1.***

VOICES FROM AROUND THE TABLE

—

A recent (March 13, 2016) article in the *Washington Post* begins with this sentence: "Finn Laursen believes millions of American children are no longer learning right from wrong, in part because public schools have been stripped of religion." (Incidentally, by *religion*, it is evident that Mr. Laursen means *Christianity*.)

Do I *dare* comment? To do so only invites misinterpretation. But there are just *so many* hidden issues begging for exploration here. There are so many other different voices to be heard, as it were.

Let's just sit back and listen to some of those voices from around the "table of common goals." Have tea with any of them, and explore further, if you wish.

A disclaimer: This column is *not* about religion, pro or con. That would miss the point. It would miss several points, actually. This is solely about *education* in general.

Back to the voices arising from around the table. These voices are not necessarily mine, though I find points of agreement with each. None of these voices, at least as I'm hearing them, are spoken with anger, but instead with concern and compassion for our common goals. Each voice addresses a different relevant issue/concern raised by the opening quote. We'll miss some voices, but we'll hear as many as we have space for.

> **VOICE 1:** Our schools have not been "stripped" of religion, for that would require that "religion" be in schools in the first place. That's *not* meant to be negative or controversial at all. It's meant to remind us that *public* schools are provided/established by the "state," and our founding fathers and framers of the Constitution *specifically intended* that church and state be separate. This extended to public schools, precisely to avoid the potential for the kind of "state religion" or indoctrination that many of our founders' heirs had come to America to escape.

Our founders were wise enough to know that mixing *any* religion in government or in churches, *or* in schools, is dangerous. This point is absolutely crucial. (See also *Education and the Founding Fathers* earlier in this Section, especially Point #4.)

VOICE 2: Treading carefully here, but do we *want* our public schools to have the burden of "teaching" right & wrong across the board? (Especially when even religion itself disagrees on bigger issues like abortion, same-sex marriage, and others?) Aren't many of our *basic values* decisions the purview of parents, supported perhaps by society, and indeed, perhaps by religion itself? We can't continue to add more societal expectations to our schools (and strip them from our parents?), and then complain if they don't do it the way we think they should. This is not fair to our schools, or to us.

VOICE 3: We have another example here of the actual [delightful] "fuzziness" of "education." This has been an underlying theme of every one of these columns. All of us are in favor of a "good education" but it seems that no two definitions of that are the same. *What* do we want of our schools?

VOICE 4: This also effectively highlights the very real difference between *teaching* and *learning.* We may not want our teachers to be *teaching* (weighing in on) specific questions of right/wrong. On the other hand, our students often *learn* by watching. When they see a teacher or parent or adult or politician (not always the same as "adult") doing the right—or even wrong—thing, they are influenced, and they *learn.* We all want our teachers doing "the right thing," and often, in these situations, that is the better way for students to *learn.*

I still hear *several* more voices wanting to speak, but time & space require that they wait. In the meantime, we have lots of food for thought.

Epilogue

When I was a young faculty member—full of energy and setting out to conquer the world—I swore I would never be one of those "old farts" complaining about how students had changed, how much less students knew now than when these colleagues started teaching, and, to my vain, youthful ears—a litany of other tired complaints.

Change is the name of the game, isn't it?

Now I'm a retired faculty member (older now, than those colleagues were then), past my prime, and looking back at a career that seems like a thirty-minute movie. I don't *think* I ever developed the perceived attitude I questioned back then, but I recognize I may have been a little naïve. And I recognize my predecessors may have been saying more than I thought.

Of course, students change—then and now. Times change. Today's students are "millennials" (or even children of millennials!), a far cry from Baby Boomers, or Gen X or Y'ers that are their parents and grandparents. And all of us have weathered lots of changes, as has the world.

Nothing new here. We all know this, at least intellectually. As the pundits say, the only constant *is* change. And, to use an extreme example, we can't use one-room schoolhouse techniques on children who barely even *know* of *Little House on the Prairie* lives. We know this intellectually as well, but the faster things change, the more complicated education gets, and the more we need to be reminded—almost daily.

In education, this constant-change reality creates constant (and changing) challenges as well as opportunities, of course. But dealing effectively with changes in education can be as difficult as boxing a cloud. Changes and trends around us can be hard enough to identify as they occur, let alone to (try to) keep up with. It's a constant feeling of being behind, in a system that doesn't allow quick changes.

But to be effective in the long run, education must do even more: it must prepare students for the unseen (and even unforeseen) changes down the road. We used to have a professor that said, "Our job is to prepare you for jobs that don't exist yet."

It's a tall order. No matter how good the system is (and it is aging, too), *we are always trying to prepare students for the future, in classrooms of the present, with tools and perspectives that are frequently stuck in the past.*

And there's one other thing to note. Not only does education have to keep up with—and prepare for—the changing nature of the world, it must keep re-defining and re-inventing itself, almost on the fly, to do that. It's the nature of the game, and it's very complicated.

As I look back on my predecessors' complaints, I'm not sure they were so terribly bothered by the changing nature of students and the world, after all. And I'm not even sure the frustration was *with* the students. I think I was too young to recognize and articulate this, but I think there's a good chance that they were expressing a vague lament that the educational system hadn't kept up with the changing world, and their students had suffered for it.

Or maybe not. Maybe they *were* just old curmudgeons, after all. But, looking back now, I like to think they knew better. I prefer to think they knew—as I didn't quite know yet—that keeping education abreast of the rapidly changing times is never *ever* easy.

Forewarned is forearmed, as they say. And so, we come to the end of this collection of thoughts by looking the only way we can look: to the future. If we all work together, and stay aware, and remain creative, and *listen* to each other, we can keep "keeping up." We have no other choice. Our future depends on it.

APPENDICES

NOTE:

Quicker access to the links mentioned in these appendices may be gained by visiting an electronic version of all 3 of these Appendices. Use the following URL in your browser: https://aftermathenterprises.com/appendices-rollin-river/

APPENDIX 1: RESOURCES AND MORE INFORMATION

—

1. In Section 1, p. 24, the footnote mentions a resource used to pattern the "quiz" in that essay. Here is a link to the original book, from National Council of Teachers of Mathematics. It is the best book on the general topic I've seen. It is not only readable, but provides a wonderful perspective and things to ponder.

http://www.nctm.org/store/Products/Mathematics-Assessment--Myths,-Models,-Good-Questions,-and-Practical-Suggestions/

2. In the Education in an Information Age essay (Section 3, p. 64), there is a Doonesbury cartoon mentioned. It's amazing to me how successfully that particular strip captures the essence and perspective of this topic, while simultaneously mixing in humor. It is perhaps my favorite "educational topic" cartoon strip. The link is provided in the piece, but here it is again:

http://www.gocomics.com/doonesbury/2011/06/26

3. Notwithstanding the cartoon strip mentioned above, there is/was no cartoon strip I know of that matches the sheer volume of excellent-while-humorous educational (and mathematics educational) cartoon strips than **Calvin and Hobbes**. Again, the humor helps drive home the thought-provoking points. Check out this site for a large collection of samples:

https://www.google.com/search?q=calvin+and+hobbes+education&tbm=isch&tbo=u&source=univ&sa=X&bed=oahUKEwi1pOiWr8rUAhUn4IMKHTDyC2UQsAQIJQ&biw=1366&bih=589

4. As to "the answers" to the Poetry/Mathematics quiz at the end of Section 4 (p. 89-92), here is the quick version: All the odd-

numbered quotes refer to poets/poetry, while all the even-numbered ones refer to mathematicians/mathematics. For the original version of the article, check out the link below. (This is a long [and perhaps surprisingly "academic"] article, but the quiz and answers are given in Appendix B at the end.)

http://scholarship.claremont.edu/cgi/viewcontent.cgi?article=1428&context=hmnj

5. The selection "**Notes From the River...**" in Section 10 mentions a trip I took in Summer 2016. For seven weeks (51 days) that summer, I explored (by car) the entire Missouri River from its official source in Three Forks, MT all the way to where it empties into the Mississippi north of St. Louis, MO. What an amazing adventure! Growing out of that trip came my first book **Rollin' Down the River: Discovering People and Places Along the Mighty Missouri.** (Acclaim Press)

6. **Further/ongoing resources.** Preferably, this section can take on an ongoing and growing nature through the author's website(s). If readers have questions or comments, feel free to use the connections and information detailed in Appendix 2. Resources that grow out of those enquiries and other places will be added to the list, which can be periodically checked (see LarryNCampbell.com). Feel free to check back often.

APPENDIX 2: FURTHER CONNECTIONS WITH AUTHOR

Larry Campbell currently resides in Branson, MO, and he would be delighted to hear from you. He may be reached at larrycampbell@missouristate.edu or larrycampbell@AfterMathEnterprises.com.

To visit Larry's general **AfterMath Enterprises** website, go to www.AfterMathEnterprises.com. While at this site, you can also sign up for his (free) Bi-Weekly Photo/Sharings, delivered on (usually alternate) Monday mornings. These always contain one of his photos from around the world (including his 2016 Missouri River trip), along with other features and "Wild Cards" of multiple varieties, all designed to brighten your Monday mornings—and your week.

To visit Larry's author website, go to www.LarryNCampbell.com. This site will contain other book-related connections (including an ongoing Resources link, mentioned in **Appendix 1**).

Larry is available for Book Signing Events and/or presentations on a variety of subjects, and he enjoys doing these events. Topics are varied. Some are similar in nature to those in this book, including a talk on "Interesting and Weird Mathematicians and Stores about Them". He also speaks on his 2016 River Adventure all the way down the Missouri River (by car) from source to mouth, and has several "travelogue" type presentations, with pictures of his various world travels. For further details, and/or to compare calendars, contact him at the address(es) above.

APPENDIX 3: ASSORTED BRAIN TEASERS

—

Also known as "Yoga for the Mind"

"Brain Teasers" are wonderful tools for the budding problem solver in mathematics and elsewhere, but they are ***not*** For Math Nerds Only! They can be a great deal of fun, as well. Listed below is a sort of "Top Ten" of some Easy-To-State Favorites, after which some links will be provided to others, as well. **Lighten Up. Pick (only) the ones you that grab you, and by all means, *have fun!***

—

1. The Horse Trader

A man buys a horse for $60, sells it for $70, buys it back for $80, and finally sells it for $90. Did the make any money on the transactions, and if so, how much?

2. Handshakes

Ten people enter a room for a meeting. If they all shake hands with each other (once each), how many total handshakes will there be?

3. Dollars and "Sense"

In my hand, I have two common US coins whose total value is 55 cents. One of the coins is not a nickel. What are the coins?

4. Calendar Cubes

How are the digits 0 – 9 placed on two separate cubes to form a calendar (often seen in banks, etc.) that can display every date of any given month? (Always use both cubes: 01, 02, 03,... 30, 31)

5. It's All "Relative"

"Brothers and sisters have I none, but this man's father is my father's son." To whom is the speaker referring?

6. Buckets of Water

You have TWO completely unmarked buckets. One will hold exactly 7 gallons, the other 4 gallons. How can you measure exactly ONE gallon of water? (What if the smaller bucket holds 5 gallons instead?)

7. Number Sequence?

What are the next two letters in this sequence?

O, T, T, F, F, S, S, ____, ____

8. The Frog in the Well

A frog is at the bottom of an empty 30 foot well. Each day he is able to climb up 5 more feet, but at night, he slips back 3 feet. At this rate, how long will it take him to escape the well?

9. The "Average" Driver

Ralph drives 60 miles at an average speed of 30 mph, then returns over the same route at an average speed of 60 mph. What was his average speed for the round trip?

10. Word Trivia

Consider the *spellings* of each positive whole number. **A**) Which number's spelling is the first to contain the letter "a?" **B**) Which number's spelling is the *only one* in alphabetical order? **C**) Which is the only one in *reverse* alphabetical order?

—

ANSWERS and More Brain Teasers?

Feel free to share proposed solutions and/or questions with the author, using the links provided in Appendix 2. In order to promote the needed "play/think" time for these, the answers are not immediately given in a link. Contact the author for a link to the answers.

To see an archive of Brain Teasers *and answers* that have been used during a roughly 2-year span of Weekly Mailings, visit http://aftermathenterprises.com/various-answers/. There are also other links to more Brain Teasers available by request.

If you want to receive periodic Brain Teasers (followed by answers later), visit the website and sign up for the Weekly (sometimes Bi-Weekly) Photo Sharing, in which Brain Teasers are usually included.

Dr. Larry Campbell spent most of his professional career working as a professor in mathematics and mathematics education, split equally (17 years each) between the College of the Ozarks near Branson, MO and Missouri State University in Springfield, MO. He also spent three years between those stops as President of Ozark Mountain Community Classroom in Branson.

He retired from Missouri State — in stages — between 2010 and 2012, and, since then he has been running AfterMath Enterprises, LLC, an umbrella organization for all the activities in which he is engaged. Besides doing talks, programs, & workshops for civic/community groups and schools, he also puts out a (free) Photo/Sharing e-mail blog which combines his photography hobby with several other Monday morning brighteners & tidbits for the week.

He currently writes a bi-weekly educational column for the *Springfield News-Leader* in Springfield, MO. This book is a collection of some of those columns.

Campbell's first book, *Rollin' Down the River: Discovering People and Places Along the Mighty Missouri* [Acclaim Press] was released in July 2017. It details his 2016 seven-week journey, following the Missouri River (by car) from its source in Three Forks, MT, to where it ends near St. Louis, MO.

Larry and his wife Pat live in Branson, MO. They have two grown children, Christi and Adam.

CPSIA information can be obtained
at www.ICGtesting.com
Printed in the USA
BVHW07s0341041018
529258BV00001B/14/P

9 781633 734685